Preface

There is currently great interest in the biology of soil, stimulated by an increased concern to conserve natural resources, not only in agriculture but also in natural ecosystems. Recent developments in molecular biology have increased the possibilities of manipulating soil organisms and the processes they carry out, to improve food production and the quality of the environment. But important environmental questions associated with the application of this new biotechnology to soils are still to be resolved. Issues such as the desirability of using genetically-engineered microorganisms in soil can be discussed sensibly only in the light of an understanding of naturally-occurring organisms in soil.

Although on a global basis food production continues to increase, man's current relationship with the soil in many parts of the world is characterised by decreasing soil fertility and increasing population pressure. Soil Science has an important role to play in the future, helping to solve these problems. Although not all areas of this large and expanding subject are covered in this book, I hope that I have been able to present the principles of soil biology in such a way that will interest the reader.

It has been my aim to produce an integrated account of the organisms in soil—from viruses to trees—the environment in which they live, and the ways in which they can modify soil by their biochemical processes; the ways in which organisms interact, particularly with plant roots, and the possibilities for manipulating soil organisms are also discussed. Only a small number of references (mainly review articles and books) have been given. These have been chosen carefully to provide a bridge which will carry readers into the detailed literature.

This book has been written for final year undergraduate and postgraduate students in agriculture, biology, botany, ecology, microbiology and soil science. Although most students will already have been introduced to biology, the *soils* aspect of this subject may be new to them. Terms which are likely to be unfamiliar to readers are briefly explained, and references to general texts on soil science are given; *Russell's Soil Conditions and Plant Growth* is particularly valuable source of detailed information.

All data have been expressed on a m^2 basis; readers are reminded that $10^4\,m^2 = 1\,ha$ and $1\,g\,m^{-2} = 10\,kg\,ha^{-1}$.

I am very grateful to the following colleagues who have spared the time and effort to read early versions of this manuscript and whose comments have been invaluable: John Dalrymple, Peter Harris, David Jenkinson, Maxine Opperman, Donald Payne, David Rowell and Alan Wild.

Finally, this book would not have been written without the support and encouragement of Diane, Rebecca and Jennifer.

MW

To my parents

Contents

CHAPTER ONE
SOILS AND ORGANISMS

1.1 Introduction

Many scientists who are attracted to studying the biology of soil find themselves at some time or other despairing over the complexity and variability of the material with which they have chosen to work. Spatial and temporal variability cause problems for all soil scientists; however, these problems are exacerbated for biologists who are faced with the inherent variability of biological systems, together with the fact that biological processes often operate at the level of the microsite. Microsites, localised points of heterogeneity generally beyond the resolution of current analytical techniques, are often invoked to explain discrepancies which arise between observations made in the laboratory and those made under field conditions. To understand the heterogeneity, the complexity and the stability of soil ecosystems is a challenge which requires knowledge of soil biology together with an appreciation of the chemical and physical aspects of soils.

This chapter considers the major features of soil as an environment for organisms, the types of organisms present in soil and their activities.

1.2 The soil environment

Soil may be studied at a range of different scales. On a global scale, the encroachment of deserts into agricultural areas of the earth's surface is a major concern (section 5.5). However, soils form only a relatively thin layer over one third of the surface of this planet, and the roots of crop plants do not grow much beyond 1 m depth of soil. The growth of a plant depends upon many factors including the way soil solids are arranged to provide channels approximately 0.2 mm in diameter (section 4.6). The supply of nutrients such as nitrate to a plant depends upon the activity of micro-

Table 1.1 The size of some soil components.

Component	Diameter (μm)
Sand particle	2000–20
Plant root	200
Transmission pore	> 50
Ciliate	20
Silt particle	20–2
Root hair	10
Bacterium	1
Storage pore	50–0.5
Clay particle	< 2
Residual pore	< 0.5
Aluminium atom	5×10^{-5}

organisms such as bacteria which are approximately 1 μm in length (section 1.3.2.2). The survival of such bacteria in an acid soil may depend upon their ability to exclude or detoxify aluminium atoms 0.05 nm in diameter (section 4.3.1). An awareness of these different scales (Table 1.1) is important when trying to understand the biology of soils.

Soil is a mixture of *inorganic material* (sand, silt and clay particles), non-living *organic matter* and living *organisms* (biomass) with the particles arranged into solid structures with spaces between them which contain air and soil solution. In the mid-nineteenth century soil was considered simply as a mixture of disintegrated rock and decomposing organic matter. In the latter part of that century Russian scientists, notably Dokuchaev, observed that soils consist of distinct *profiles* of related *horizons*.

1.2.1 Inorganic material

Sand particles (20–2000 μm diameter) and *silt* particles (2–20 μm diameter) are mostly quartz, and are relatively inert. *Clay* particles are composed of clay minerals and often other minerals such as quartz, calcium carbonate and oxides of iron and aluminium. Clay minerals occur in soils as primary minerals or as weathered or re-synthesised secondary minerals. They comprise mainly aluminium silicates arranged in plates or lamellae 0.7–1.0 nm thick. The lamellae are stacked together to form *clay crystals* and these crystals may become oriented along the surface of pores or sand grains, forming clay skins in certain soils. A thin section of soil, when viewed under a petrological microscope, may show regions of oriented clay lamellae (Figure 1.2), which are often larger than clay particles, termed *clay*

Table 1.2 Surface area and cation exchange capacity (CEC) of soil particles. After White (1987).

Particle	Specific surface $(m^2 g^{-1})$	CEC $(meq\ 100 g^{-1})$
Sand	0.01–0.1	0
Silt	1.0	0
Clays:		
Kaolinite	5–100	3–20
Montmorillonite	700–800	100
Vermiculite	300–500	100–150

domains. The proportion of sand, silt and clay determines the *texture* of a soil.

The size and shape of the clay minerals confer on them a large surface area, most pronounced with the smectite clays, montmorillonite being the most common example (Table 1.2). Furthermore, the substitution of one atom within a clay lamella by another atom of similar size but different charge (*isomorphous substitution*), for example aluminium for silicon, gives rise to a net negative charge in the clay particle. This charge is independent of the pH of the surrounding soil solution. However, the edges of clays and the surfaces of hydrated oxides of iron and aluminium have a charge which depends upon pH. As the acidity increases, the surface charge becomes more positive. Carboxylic and phenolic groups in soil organic matter also contribute pH-dependent charge (becoming increasingly negative at pH values > 3) of 150–300 meq $100 g^{-1}$ organic matter.

The predominantly negative charge on the surface of clays and organic matter attracts positively-charged cations. The cations which are adsorbed to these surfaces are freely exchangeable with cations in solution and this property, termed *cation exchange* (Table 1.2), is important in determining the concentration and movement of nutrients in soil. Clay minerals have a range of effects on microbial processes in soils (section 4.9).

1.2.2 *Organic matter*

Soil organic matter is derived from the organisms which live in or on soil. The variety of organisms, ranging from bacteria to trees, and the vast number of molecules making up these organisms, ranging from simple amino acids to complex polymers such as lignin, together with their decay products, give rise to a highly complex material. The precise nature of soil organic matter remains unclear.

The surface layers of a soil comprise partially-decomposed material, mainly of plant origin. Beneath this litter is a layer of amorphous brown material in which the identity of the original material is lost. This material, termed *humus*, is very poorly understood, for example only approximately 50% of the organic nitrogen in soil has been identified as compounds such as amino acids, amino sugars and nucleic acids. Humus appears to consist of resistant plant material, together with new compounds synthesised from the breakdown products of less resistant material.

Chemical extraction and analysis of soil organic matter have yielded fractions such as humic acid and fulvic acid, but the biological significance of these fractions is uncertain. Other attempts to understand the composition of organic matter have classified fractions according to their rates of decomposition (section 2.2.2). The simplest approach is to consider organic matter as a mixture of labile and resistant substrates for organisms. More complex models subdivide these fractions further to include, for example, physically and chemically protected fractions (section 5.3).

1.2.3 *Soil structure*

The arrangement of sand, silt, clay and organic matter in soil gives the soil *structure* (Figure 1.1). Interparticle forces hold clay particles together and also bind them to sand and silt surfaces. Organic matter and hydrated oxides of iron and aluminium assist in stabilising these groups of particles termed *aggregates* (section 5.4.1). Figure 1.2 shows a schematic arrangement for soil as seen at two levels of magnification.

The size and arrangement of the solids determine the size and shape of the spaces (*pores*) between them. The pores in a soil determine the air and water relationships of that soil, and therefore affect biological activity (section 4.8). Pores, which are joined to one another by narrow channels or necks between particles, can be grouped in order of decreasing size into transmission pores, storages pores, and residual pores (Figure 1.2). The pores form a continuous three-dimensional network throughout the soil (Figure 1.1). The significance of these different pores in relation to soil water is discussed in section 4.6.

Figure 1.1 illustrates the complex physical nature of the environment in which soil organisms live, and into which micro-organisms are often introduced (section 6.3). Properties of soils such as temperature, salinity,

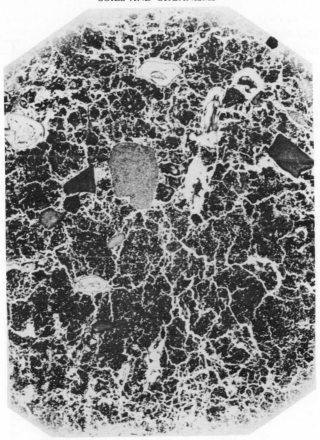

Figure 1.1 A thin vertical section of the top 9 cm of a permanent pasture soil (Linhope Series) from Woodburn, Southern Uplands, Scotland. (Courtesy of R.J. MacEwan, University of Reading.)

acidity, moisture and aeration, which are important factors in the ecology of soil organisms, are discussed in Chapter 4.

1.2.4 *Soil classification*

The formation and development of soils is due to a complex interaction of chemical, physical and biological factors (Chapter 5). Although soils form a

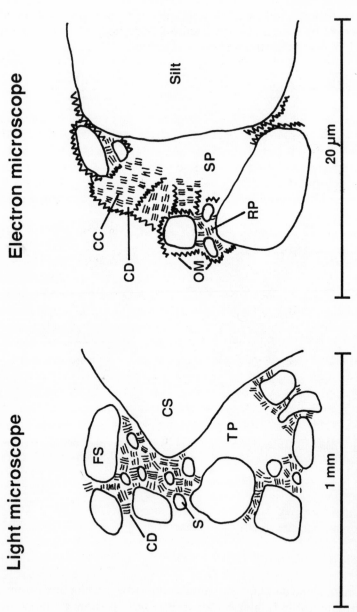

Figure 1.2 A schematic diagram of soil particles as seen at two levels of magnification (*CC* clay crystal, *CD* clay domain, *CS* coarse sand, *FS* fine sand, *OM* organic matter, *RP* residual pore, *S* silt, *SP* storage pore, *TP* transmission pore). (Courtesy of Dr D.L. Rowell, University of Reading.)

continuum of variation they may be classified according to the degree of similarity between individual soil profiles. For example, *podzols* are characterised by strongly bleached subsoil horizons, and *chernozems* are soils rich in organic matter and black in colour.

1.3 The soil population

The organisms inhabiting the soil include micro-organisms, plants and animals. A knowledge of the organisms present is of some interest, but the presence of living organisms leads to biochemical changes in soil, and in order to understand the way in which soil functions, information is needed on these activities. These include the reactions carried out by organisms, the interactions which occur between organisms and between organisms and their environment.

Studies on soil organisms may be made at a range of different levels (Table 1.3). Scientists interested in the cycling of nutrients within a particular ecosystem may only need to know the amounts of plant biomass or microbial biomass in soil (section 1.3.1). Ecologists studying the food web in a soil–plant system (section 1.4.1) are concerned with the relationships between different functional groups such as bacteria and bacterivorous nematodes (section 1.3.2). Those interested in the introduction of a genetically engineered bacterium into soil may need to know the fate of a single strain of a single species or even a single organism (section 6.4).

Alternatively, scientists concerned with processes in soil, for example the contribution of carbon dioxide derived from soil to the concentration of carbon dioxide in the atmosphere, may only be interested in the total amount of respiration by soil. Those attempting to measure the efficiency of utilisation of fertiliser nitrogen by crops must consider a group of processes,

Table 1.3 Levels at which soil biology may be studied.

Level	Example
Organism	
Population	Microbial biomass
Functional group	Bacterivorous nematodes
Species	*Pseudomonas fluorescens*
Activity	
Total	Soil respiration
Process	Fertiliser recovery
Enzyme	Nitrogenase

including nitrification and denitrification (sections 2.4.2 and 2.4.3). Scientists trying to understand the biochemical basis of the ability of legumes to fix atmospheric nitrogen are concerned with a single enzyme, nitrogenase (section 2.4.1). Such approaches are complementary, and as more information is obtained at these different levels of study, an understanding of soil biology develops.

1.3.1 *Biomass*

Most soils are colonised by plants, and soils are generally considered in terms of their ability to support plant growth. However, the presence of plants (*primary producers*) leads to an input of organic material into soil, and it is this material that supports most of the remaining soil population. The fixation of carbon dioxide during photosynthesis provides the carbon and energy for life in the soil. However the *net primary production* (the rate of production of biomass after allowing for losses due to respiration) varies in different ecosystems (Table 1.4). Production is greatest in tropical rain forests (up to $3500 \, g \, m^{-2} \, yr^{-1}$) and decreases as photosynthesis is reduced by extremes of temperature and moisture.

These differences in primary production lead to differences in total organic carbon in soils (Table 1.5) and in *microbial biomass* (the living part of the soil organic matter, excluding plant roots and soil animals larger than approximately $5 \times 10^3 \, \mu m^3$). Comparison of Table 1.4 and Table 1.5 shows that although the tropical rain forest has the highest primary production, the soil that supports this vegetation contains a relatively low amount of organic matter. This is due to the rapid decomposition of plant litter in this ecosystem (section 5.3). Soil under temperate permanent grassland con-

Table 1.4 Net primary production in different ecosystems. After R.H. Whittaker and C.E. Likens (1975) in *Primary Productivity of the Biosphere*, eds. H. Leith and R.H. Whittaker, Springer-Verlag, 305–328.

Ecosystem	Normal range $(g \, m^{-2} \, yr^{-1})$	Mean $(g \, m^{-2} \, yr^{-1})$
Tropical rain forest	1000–3500	2200
Temperate deciduous forest	600–2500	1200
Boreal forest	400–2000	800
Savanna	200–2000	900
Temperate grassland	200–1500	600
Tundra and alpine	10–400	140
Desert and semidesert scrub	10–250	90

Table 1.5 Organic matter and microbial biomass contents (expressed as carbon) for soils from different ecosystems. After Jenkinson and Ladd (1981).

Ecosystem	Depth (cm)	Organic C $(g\,m^{-2})$	Biomass C $(g\,m^{-2})$	Biomass C as % organic C
Tropical rain forest	0–15	1900	76	4.0
Temperate grassland	0–23	7000	224	3.2
Temperate arable	0–23	2900	66	2.2

tains more organic matter due to the dense root mass in grassland soils and the slower rate of decomposition. Values for microbial biomass in these soils reflect the organic matter content, an indication of the amount of *substrate* available for maintenance and growth (section 2.1.2). The values for microbial biomass form a fairly constant proportion of the total organic carbon, ranging from 2–4% (Table 1.5).

1.3.2 *Functional groups*

There have been only a small number of studies which have quantified the major groups of organisms in particular soils, and only a few of these have considered the interactions which might occur between these different groups. This paucity of information is partly due to the lack of reliable techniques for quantifying the numbers and sizes of the different organisms in soil, and also because of the labour-intensive nature of the existing techniques. Some studies have identified individual species present in soil, but have not related this information to the activity of these organisms.

The role of particular organisms can be deduced in a variety of ways; this allows the organisms to be arranged into *functional groups*. Specific groups can be added to sterile soil (this system is termed a *microcosm*) and the effect of this addition on a process such as nitrogen mineralisation can be measured. Conversely, selective biocides may be used to eliminate specific groups of organisms such as nematodes and the effect of this removal measured. Alternatively, the dynamics of major groups can be followed in soil and changes in particular groups correlated with particular processes. Organisms such as nematodes may also be grouped according to morphology and other biological characteristics.

Organisms may be classified according to their body size (length or width) (Table 1.6). The use of litter bags with different mesh sizes to exclude particular groups of organisms allows the role of these organisms in

Table 1.6   Classification of soil organisms according to body width. After M.J. Swift *et al.* (1979).

Grouping	Body width	Organisms
Microflora	$< 10\,\mu m$	Bacteria, fungi
Microfauna	$< 100\,\mu m$	Protozoa, nematodes
Mesofauna	$< 2\,mm$	Collembola, acari, enchytraeids, termites
Macrofauna	$< 20\,mm$	Millipedes, isopods, insects, molluscs, earthworms

decomposition of litter to be deduced. Such experiments suggest that the microfauna are rarely involved in litter comminution, the mesofauna do attack plant litter, but their overall contribution is small (termites are the exception to this) and the macrofauna have the major effect on litter decomposition. The role of earthworms and termites in soil formation and development is discussed in sections 5.3.1 and 5.3.2 respectively.

Table 1.7 shows the estimated biomass of groups of organisms in a woodland soil in the UK. This is a mixed deciduous forest in the north-west of England with an acid soil. Total net primary production for the forest has been estimated to be $1308\,g\,m^{-2}\,yr^{-1}$. This system is not at steady state; 46% of above ground production is added to the standing crop, the remainder ($546\,g\,m^{-2}\,yr^{-1}$) enters the soil as litter or leachates. This acid

Table 1.7   Biomass estimates for functional groups of organisms in a woodland soil (Meathop, UK). After J.E. Satchell (1971) in *Productivity of Forest Ecosystems*, Unesco, Paris, 619–629, and T.R.G. Gray *et al.* (1974) *Revue d'Ecologie et de Biologie du Sol* **11**, 15–26.

Functional group	Biomass ($g\,m^{-2}$)
Bacteria	3.7
Fungi	45.4
Protozoa	0.1
Nematodes	0.2
Enchytraeids	0.4
Earthworms	1.2
Molluscs	0.5
Acari	0.1
Collembola	0.2
Diptera	0.3
Other arthropods	0.6
(Annual litter production	764.0)

Table 1.8 Biomass estimates for functional groups of organisms in a shortgrass prairie soil (Colorado, USA). After H.W. Hunt *et al.* (1987) *Biology and Fertility of Soils* **3**, 57–68.

Functional group	Biomass $(g\,m^{-2})$
Bacteria	60.80
Saprophytic fungi	1.26
VA mycorrhizas	0.14
Amoebae	0.76
Flagellates	0.03
Phytophagous nematodes	0.06
Fungivorous nematodes	0.08
Bacterivorous nematodes	1.16
Omnivorous nematodes	0.13
Predaceous nematodes	0.22
Fungivorous mites	0.61
Nematophagous mites	0.03
Predaceous mites	0.03
Collembola	0.01

soil is dominated by fungi with a biomass of $45.4\,g\,m^{-2}$ compared to the total biomass of the soil and litter fauna of $3.6\,g\,m^{-2}$.

Data shown in Table 1.8 are for a prairie soil in the USA. In this case it has been assumed that only 10% of the total fungal hyphae are active. This soil is dominated by bacteria, and the sub-division of groups such as nematodes reflects an increased understanding of the roles of these organisms (section 1.4.1). Ciliates, earthworms, termites and insect larvae were not major components of this particular ecosystem.

Studies have often concentrated on bacteria and fungi, which form the largest groups in terms of biomass in the soil, without considering the higher trophic levels. However groups of organisms such as mites, although a small component, may be important in determining rates of nutrient cycling and the stability of the soil ecosystem (section 1.4.1).

The main functional groups in soil are now considered. Plants have already been considered in section 1.3.1 and their important roles in soil are discussed in subsequent chapters.

1.3.2.1 *Viruses.* Viruses are a group of non-cellular infectious agents. They are obligate parasites of cellular organisms and may be more correctly considered as complex molecules of protein and nucleic acid, rather than organisms. They have no metabolism of their own, they vary in shape, and the virus particle (a *virion*) ranges in diameter from 20–250 nm.

A range of plant, insect and human viruses are found in soil, and they often persist for several years. For example, plant viruses survive in soil for up to nine years in the absence of their host, and the nuclear polyhydrosis virus of the cabbage looper (*Trichoplusia ni*), although surviving only one month on the leaves, persists for up to four years in soil.

Human viruses such as picornaviruses are introduced into soil in the effluent or sludge from sewage (waste water) treatment works. Human pathogens including helminthic parasites and human enteric viruses are neither completely removed nor inactivated by conventional sewage treatment methods such as trickling filter and activated sludge techniques. Little is known about the survival of human viruses in soil, but poliovirus can survive in soil for three–four weeks. Soil conditions such as moisture and the types of adsorbing surfaces (section 4.9) are likely to be important in affecting the survival and movement of viruses in soil.

1.3.2.2 *Bacteria.* Bacteria are the smallest organisms in soil (unless viruses are considered as organisms). They are *prokaryotes* (organisms with a relatively simple cell structure when viewed under the transmission electron microscope, with no internal organelles), whereas all the other groups are *eukaryotes* (organisms with a relatively complex internal cell structure, with nuclear membrane, mitochondria and, in some organisms, chloroplasts). Bacteria can be grouped according to their reaction with Gram's stain which depends upon cell-wall components; those organisms which retain the stain are termed *Gram-positive*, and those that do not retain the stain are termed *Gram-negative*.

Bacteria are generally rod or cocci shaped and up to several μm in length and approximately 10^{-12} g in weight. The numbers of bacteria in 1 g of soil vary from 10^6 to 10^9. Numbers estimated by the dilution plate technique detect only about 1% of those measured by direct counting. Bacteria are not uniformly distributed in soil but are located in small colonies (Figure 1.3a, c) often associated with sources of organic substrates (for example, plant roots).

One of the most important features of bacteria as a group is their biochemical versatility. An organism such as *Pseudomonas sp.* is able to metabolise a wide range of chemicals including pesticides, whereas *Nitrobacter sp.* is only able to derive energy from the oxidation of nitrite to nitrate. *Thiobacillus ferrooxidans* obtains energy from the oxidation of reduced sulphur compounds and of ferrous ions, and has an optimum pH

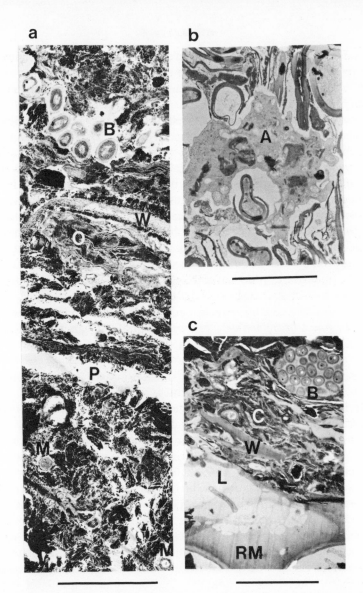

Figure 1.3 Electron micrographs of soil (scale bars 5 μm): (*a*) bacteria (*B*) associated with organic matter which still contains carbohydrate (e.g. cell walls *W*); other highly lignified and convoluted organic matter (*O*) does not support bacteria, but there are numerous micro-organisms (*M*) scattered throughout the clay; *P* is a pore about 1 μm in diameter (*b*) an amoeba (*A*) in an organic-rich surface soil (*c*) root surface mucilage (*RM*) which has been partially lysed (*L*) by soil bacteria; the mucilage holds cell remnants (*W*), clay particles (*C*) and a colony of bacteria (*B*) onto the root surface. From R.C. Foster (1985) *Quaestiones Entomologicae* **21**, 609–633.

value for growth of 2. *Clostridium sp.* is able to grow in the absence of oxygen and can obtain nitrogen by the reduction of atmospheric nitrogen gas. *Rhizobium sp.* form N_2-fixing nodules on the roots of leguminous plants (section 3.4).

Of the 190 genera of bacteria listed in the seventh edition of *Bergey's Manual of Determinative Bacteriology*, 97 genera (57%) contained species that may be considered as soil bacteria. Corynebacteria, for example *Arthrobacter sp.*, form at least half of the total bacterial colonies which grow on dilution plates. Spore-forming bacilli and actinomycetes are also common. The genera mentioned in the previous paragraph, which bring about many of the major biochemical transformations in soil, probably constitute less than 10% of the total bacterial population. Other bacteria include actinomycetes, which produce filaments or mycelia similar to fungi, but they are smaller in diameter. In culture media *Streptomyces sp.* and other actinomycetes produce antibiotics, and soils are the main source of organisms which are used to produce these compounds commercially.

Soils may also contain some bacteria which are pathogenic to humans. For example, *Clostridium spp.* are common in soil and animal faeces. They are anaerobic organisms which produce spores, and some species such as *C. tetani* and *C. perfringens* can enter wounds and cause the diseases tetanus and gas gangrene respectively. These diseases may be lethal; prophylactic immunisation is used to prevent tetanus.

Bacillus anthracis is the only species of *Bacillus*, a common soil genus, which is pathogenic to humans. Like all bacilli it produces spores which, in this case, cause the disease anthrax, and which may persist in soil for many years. In the 1940s the island of Gruinard off the north coast of Scotland was infected with spores of *B. anthracis* as part of a biological weapons test. The persistence of these spores in the soil meant that the island was uninhabitable until 1987 when the spores were eradicated by treating the soil with formaldehyde.

An early attempt at a functional classification for soil bacteria drew a distinction between those bacteria which exist mainly in a resting state with brief periods of activity when substrates are available (termed *zymogenous*), and those which are continuously active but at a low level (termed *autochthonous*) (section 2.1.2).

1.3.2.3   *Fungi.* Fungi produce filamentous *mycelia* composed of individual *hyphae*, 0.5–10 μm in diameter, which may or may not be sub-divided by crosswalls (*septa*). Fungi are less biochemically versatile than bacteria, for example they are all aerobic heterotrophs (section 1.4), but they are more variable in their morphology, for example in the range of fruiting bodies they produce. 1–75% of total hyphae are active depending on soil conditions, and the hyphae are often irregular in shape and size due to close contact with soil particles.

There are many lists of genera of fungi isolated from soil, but most of this information is obtained from the dilution plate technique. However, many soil fungi, such as the larger Ascomycetes and Basidiomycetes, do not grow or sporulate readily on soil dilution plates. Over 690 species of fungi from 170 genera have been isolated from soil, with a small number of genera, such as *Penicillium* and *Aspergillus*, accounting for more than half of the species.

Fungi produce a range of specialised structures in soil. Thickened strands of mycelia termed *rhizomorphs* are often conspicuous in the litter layers of woodland soils. Thick-walled structures such as *chlamydospores* (vegetative resting spores), *vesicles* containing oil globules, and *sclerotia* which consist of closely interwoven hyphae 20–30 cm in length, are widespread. Fungi also produce a variety of spores which appear to have only limited survival ability. A wide range of fruiting bodies are formed which vary in size and complexity. Most of the larger fruiting bodies produced by Ascomycetes and Basidiomycetes occur at the soil surface, for example those of agarics found in woodland and pastures.

Some fungi are also important plant pathogens (section 3.2), for example *Gaeumannomyces graminis*, the causative agent of take-all disease in wheat, which overwinters in soil on debris from the previous crop ready to infect the following crop. Other fungi form a symbiosis with plant roots, termed a mycorrhiza, which is important in improving the supply of phosphorus to the plant (section 3.7).

1.3.2.4 *Cyanobacteria and algae.* Cyanobacteria, formerly known as blue-green algae, and green algae (chlorophyceae) are photosynthetic organisms ubiquitous in soils. Data for a range of virgin soils and cultivated chernozems in the USSR indicate values for cyanobacterial and algal biomass of 0.7–54.6 g m^{-2}. They are usually most abundant close to the soil surface, but may also be found at depths down to 15–20 cm below the surface, where there is insufficient light for photosynthesis. The contribution of these organisms to net primary production is not clear.

1.3.2.5 *Protozoa.* Soil protozoa consist of *flagellates* (mean diameter 5 μm), for example *Bodo sp.*; *amoebae* or *rhizopods* (mean diameter 10 μm), for example *Euglypha sp.*; and *ciliates* (mean diameter 20 μm), for example *Colpoda sp.* Amoebae (Figure 1.3 b) either consist of a mass of protoplasm surrounded only by a flexible pellicle (*naked amoebae*), or possess a rigid shell made of chitin or silica with fine protoplasmic strands emerging from the end (*testaceans*). Flagellates have one or more flagella attached to one end of their bodies which allow movement, whereas ciliates use cilia which are often arranged in bands.

There is little field information on protozoa, partly due to the difficulty in studying these micro-organisms. They lack a proper cell wall, and changes in pH and salt concentration, or mechanical damage, can cause the cells to burst. Furthermore, they are small and variable in shape, and because their numbers are at least an order of magnitude less than for bacteria, direct counting is impossible. The biomass of protozoa in an arable soil can be almost equal to that of the earthworm population, but protozoa have a much shorter turnover time (section 5.3). A Rothamsted soil contained 70 500 flagellates g^{-1}, 1400 amoebae g^{-1}, and 377 ciliates g^{-1}. Although data often indicate that flagellates and amoebae are more numerous than ciliates, when the mass of the individual organisms is taken into account (0.2–28.0 ng for flagellates, 0.8–6.0 ng for amoebae, and 1.5–750 ng for ciliates) then the ciliate biomass may equal or exceed that of the other two groups.

Protozoa are major consumers of bacteria, and they are able to survive adverse conditions by *cyst* formation. They have been recovered from air-dry soil after being stored for 49 years. Naked amoebae appear well suited to soil because their sliding motion on surfaces enables them to feed on soil particles and roots, and their flexible cells are adapted to feeding within thin water films around soil particles (section 4.9). Many flagellates are saprozoic, feeding on dissolved organic nutrients. Ciliates, being larger organisms, are probably restricted to feeding during periods of high water content (section 4.6).

A neglected and unusual group of soil organisms with some characteristics in common with protozoa are the *myxomycetes*. They are found in a wide range of soils from Albanian woodland to the Sonoran desert and possess a combination of plant, microbe and animal features. The germination of spores produces *myxamoeba* or *flagellate cells* which give rise to *swarmers* and *swarm cells* (10–20 μm long). Myxomycete biomass has been estimated as 0.006–0.064 g m^{-2}. Little is known of the ecology of these organisms although the dung of herbivorous and omnivorous mammals provides substrates for their growth. They consume other micro-organisms and are eaten by a range of organisms.

1.3.2.6   *Nematodes*. Nematodes are a diverse group of roundworms widely distributed in soils, and when adult may be 0.5–1.5 mm long and 10–30 μm diameter. They depend upon a thin film of water around soil particles for movement (section 4.7) and completion of their life cycle; growth is better in coarse-textured soils than in fine-textured soils. However, their microscopic size, clumped distribution and variety of feeding habits present problems when investigating their role in soil. Their distribution in soil generally follows that of organic matter.

Nematodes can be sub-divided into functional groups. Some nematodes, for example *Meloidogyne sp.*, are plant parasites (section 3.8), but the majority (70%) do not feed directly on plant roots but consume other soil organisms. However, nematodes may indirectly affect plants by feeding on fungal pathogens or increasing mineralisation of nutrients (section 1.4). For example, the *fungivorous* nematode *Aphelenchus avenae* decreases the population of the fungal pathogen *Rhizoctonia solani*. *Bacterivorous* nematodes can survive on bacteria such as *Agrobacterium tumefaciens* and *Erwinia carotovora*, and 30–60% of bacteria engulfed may be defecated in a viable condition. *Omnivorous* nematodes feed on all trophic levels from root hairs, bacteria and fungi to protozoa and other nematodes.

Predatory nematodes consume protozoa, rotifers (organisms the same size as protozoa recognised by the presence of a crown of cilia), tardigrades (soft-skinned organisms less than 1 mm long with eight legs, which resemble microscopic bears), fungal spores, enchytraeids (section 1.3.2.7) and other nematodes. Nematodes are prey to certain fungi, for example *Arthrobotrys oligospora*, protozoa and other nematodes. In some species, stress conditions may induce sex reversals within individuals leading to a male-biased sex ratio, which may assist survival.

Table 1.9 Earthworm populations in a range of soils. After Edwards and Lofty (1977).

Site	Population (No. m^{-2})	Biomass (g m^{-2})
Fallow soil (USSR)	19–34	4.6–8.4
Woodland (USA)	14–142	26–280
Arable (UK)	18	1.6
Pasture (Australia)	260–640	51–152
Savannah (Ivory Coast)	18	1.7

1.3.2.7 *Earthworms.* Earthworms are found in many soils throughout the world. Populations range from < 1 to $850 \, \mathrm{m}^{-2}$ $(0.5-300 \, \mathrm{g} \, \mathrm{m}^{-2})$. Data on populations from a range of soils are given in Table 1.9. Comparisons between different soils are complicated by differences in extraction methods and in seasonal changes for a particular soil. However, there are generally fewer earthworms in acid soils (for example moorland) and bare fallow soils than in pasture soils. Numbers in arable soils are particularly variable, and there is little information about populations in tropical soils. Earthworms are not uniformly distributed in a particular soil, for example *Lumbricus rubellus* is more frequently found beneath dung pats. Different species inhabit different depths in soil, for example *Dendrobaena octaedra* lives mainly in the surface organic horizon, whereas *Lumbricus terrestris* commonly burrows down to 1 m and sometimes to 2.5 m. Subterranean species are often paler in colour than surface feeders.

Earthworms can migrate both horizontally and vertically in soil. *Allolobophora caliginosa* introduced into new polders in the Netherlands migrated horizontally at a rate of $6 \, \mathrm{m} \, \mathrm{yr}^{-1}$. In most parts of the world earthworms show a seasonal vertical migration, probably caused by the upper soil horizons becoming unsuitable for feeding and growth. Soil temperature and moisture content seem to be the major factors determining activity. For example, *Lumbricus terrestris* and *Allolobophora caliginosa* are most active in English pastures between the months of August and December and between April and May. When the temperature is above $5 \, ^{\circ}\mathrm{C}$ *Allolobophora caliginosa* is found at the soil surface, but at lower temperatures moves deeper in the soil. In the tropics most activity occurs at the start of the rainy season. In West Africa *Millsonia anomala* becomes inactive when the soil moisture content is less than 7%. Worms also exhibit diurnal patterns of activity. *Lumbricus terrestris* is most active between 6 p.m. and 6 a.m., whereas *Millsonia anomala* has two peaks of activity, one at midnight and one at 9 a.m. Their main role in soil is in the comminution of plant material prior to decomposition by micro-organisms. The role of earthworms in the development of soils is described in sections 5.3.1 and 5.4.2.

There are many reports of an increase in the numbers of micro-organisms in the earthworm gut, or in cast material, relative to the surrounding soil. The role of these micro-organisms is unclear, but it seems likely that they form an essential part of the earthworm diet, enabling the animal to grow. Most studies have been on *Eisenia foetida* (the brandling or tiger worm, found worldwide in compost heaps) because of its potential use in bioconversion of organic wastes. There is no evidence of a specialised gut microflora in *Lumbricus terrestris*. The low residence time for ingested material (20 h when feeding, 12 h when burrowing) suggests that little decomposition of resistant material occurs in the gut. There may be some breakdown of cellulose and chitin due to the production of cellulase and chitinase by the earthworm rather than by the micro-organisms in the gut.

E. *foetida* does not gain weight when feeding on mineral soil or cellulose, but does when feeding on micro-organisms. Bacteria and protozoa are preferred to fungi, and protozoa may be essential for earthworm development. Pathogenic bacteria can also be utilised. The mineral fraction of the soil also appears essential for growth. Maximum weight gain occurs on substrates with a low C:N ratio (15–35). The role of nematodes in earthworm nutrition is unclear. Field data for three New Zealand pasture soils showed that the presence of earthworms caused a 50% reduction in the total number of nematodes. Furthermore, *L. terrestris* feeding on cattle dung contains a large number of nematodes in the gut, but none in the casts.

Earthworm casts, rich in ammonia and partially digested organic matter, provide a good environment for microbial growth. Increased mineralisation in cast material relative to non-cast soil may be due to an indirect effect of worms in providing a suitable environment, or an additional direct effect in providing an inoculum.

Figure 1.4 An enchytraeid worm (approximately 2 cm long). (Courtesy of Dr S.P. Hopkin, University of Reading.)

Earthworms are extremely sensitive to soil disturbance; they cannot hear but are sensitive to vibrations. Darwin noted that worms emerging from their burrows in soil in pots took no notice of sounds from a piano until they were placed on top of the piano when they retreated at once back into their burrows.

Enchytraeids (Figure 1.4) are small worms 0.1–5.0 cm long which feed upon micro-organisms, nematodes and plant litter. Populations of approximately 200 000 m^{-2} have been found in heathland soils in Denmark and in coniferous woodland in the UK. The smaller enchytraeids may be confused with soil nematodes, but they lack the buccal stylets for piercing plants and a muscular oesophagus for sucking which are found in nematodes. The precise role of these organisms, for example *Fridericia sp.*, is unclear.

1.3.2.8   *Arthropods.* Arthropods are animals with exoskeletons and jointed legs. They include *centipedes* (Chilopoda) which are primarily carnivorous but which may also feed upon plant tissue, and *millipedes* (Diplopoda) which are vegetarian, feeding on plant material in various stages of decay. Both these groups are mainly woodland species. *Termites* (Isoptera) are important in tropical and sub-tropical regions (section 5.3.2), fulfilling a role similar to that of earthworms in temperate regions.

Beetles and wireworms (Coleoptera), midges and gnats (Diptera) and woodlice (Isopoda) are also found in soil. *Woodlice,* which may feed on dead organic matter (*saprophagous*) or on living plants (*phytophagous*), are important in soils which are periodically too dry for earthworms (section 5.3.1).

The most important members of the soil arthropod population are the *mites* (Acari) and *springtails* (Collembola). A mean arthropod population of 220 000 m^{-2} has been recorded for an old grassland soil in Ireland. Arthropods can be sub-divided into functional groups. *Predatory* arthropods either pursue prey, as do carabid beetles, or ambush prey, often injecting toxins or entangling the prey with silk.

Microphytophages feed on bacteria, fungi and algae, with fungi being the most important food. *Fungivores* may engulf their prey, and stimulate fungal growth by grazing senescent hyphae, or may pierce fungal cell walls and drain the fluid contents leaving a ghost hypha. *Detritivores* scavenge dead organic material, usually after it has been conditioned by micro-organisms. This activity by millipedes, woodlice, termites and some mites leads to comminution of organic material, stimulation of micro-organisms and deposition of faeces which can contribute to soil development.

Macrophytophages may be root sap feeders such as root aphids, or root grazers such as root weevils. Finally, a number of collembolans, mites and insects are *omnivores*, feeding on nematodes and other soft-bodied soil invertebrates. Arthropods have their greatest effect in soils dominated by fungi, for example forest soils, where millipedes, mites and collembola are dominant.

1.3.2.9 *Molluscs.* Slugs and snails (Gastropods) are the major molluscs found in soil. *Slugs* are nocturnal animals, normally active from about two hours after sunset to two hours before sunrise. A large proportion of the population of smaller slugs shelter beneath the soil surface. Slugs are a major pest in gardens, feeding mainly on plants, and can cause extensive damage to seedling crops during damp conditions. A slug population of 140 m^{-2} has been recorded in a badly damaged wheat crop. The biomass of slugs in a garden soil may be 20–45 g m^{-2}. The dusky slug *Arion subfuscus* feeds mainly on fungi and faeces of animals.

Some slugs, for example *Milax budapestensis*, ingest soil and this action, together with the production of mucus by slugs and snails when crawling, may contribute to soil structure. Faeces also contain a high proportion of partially decomposed organic matter. Slugs are consumed by toads and grass snakes.

Snails have a shell which is large enough to house them and are less destructive than slugs. Thrushes eat the polymorphic land snail *Cepea nemoralis*, and hedgehogs (*Erinaceus europaeus*) feed on many species of snails and slugs. Average populations of snails of 2 m^{-2} have been reported.

1.3.2.10 *Other animals.* A range of large animals such as rabbits, foxes, wombats and aardvarks form burrows in soil. It has been estimated that the average weight of deer, badgers, foxes, moles and voles in British woodland is 0.8 g m^{-2}.

The common mole (*Talpa europaea*) is widespread throughout Great Britain, but is absent from Ireland. The burrowing habit of the mole results in the formation of tunnel-like runs and molehills, which may be inconvenient to farmers and gardeners. Molehills may become colonised by ants which consolidate and enlarge them. A tunnel system may extend for 400–2000 m^2 and consist of deep tunnels 5–20 cm below the surface, and surface tunnels. The fore-limbs of the mole are specialised to form shovel-shaped digging organs. Moles are active almost continuously for about 4.5 h, alternating with rest periods of about 3.5 h. They feed largely on earthworms, but also eat insect larvae, slugs and earthworm cocoons. They have few predators apart from occasional tawny owls and foxes.

1.4 Organisms and nutrient cycling

The different fractions of the soil biomass serve as sinks or pools for nutrients. The interactions between the different functional groups will therefore affect the cycling of nutrients in soil and growth of plants. The transformations of elements by organisms may be viewed either in terms of the nutrition of that organism (section 2.3) or as part of the nutrient cycle for that element. Organisms can obtain their carbon from either organic compounds (these organisms are termed *heterotrophs*), or from carbon dioxide (these are termed *autotrophs*). However, autotrophs play a vital role in the carbon cycle by fixing atmospheric carbon dioxide into organic compounds which are then converted back to carbon dioxide by the heterotrophs. These transformations form the basis of the carbon cycle (Figure 1.5).

Ecologists consider nutrient cycling in terms of *food webs* in which the interactions of different functional groups bring about the transfer of energy and nutrients. This approach relies upon an understanding of the roles of different groups of organisms in soil, and the complex nature of species interactions can lead to complex food webs. Figure 1.6 shows a simple cycle, comprising three *trophic levels*, which forms the pattern of most nutrient cycles. The plants (*primary producers*) are consumed by animals

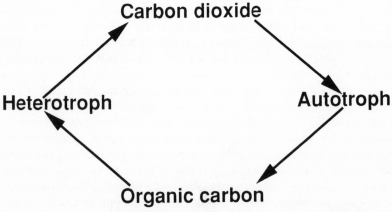

Figure 1.5 The basic carbon cycle.

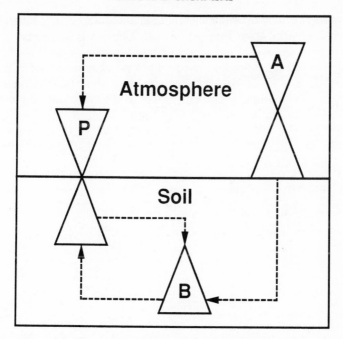

Figure 1.6 The major pools and flux pathways for nutrients in soil (*A* animals, *B* bacteria, *P* plants).

(*herbivores*) which excrete waste products and eventually die thereby providing substrates for micro-organisms (*detritivores*). Dead plant material not consumed by animals is also consumed by the microbial population. The microbial biomass plays an important role in recycling nutrients within the soil. This cycle can be described in terms of the sizes (*pools*) of the functional groups, for example the microbial biomass, and the rates of transfer (*fluxes*) of nutrients between functional groups.

The Broadbalk experiment at Rothamsted provides a relatively simple system which has been intensively studied. On one plot wheat (*Triticum aestivum*) has been grown continuously since 1843 without addition of fertiliser or manure, and the organic matter content of the soil has remained almost unchanged since 1881. The system is therefore at steady state with no net loss or gain of nutrients. The amount of nitrogen and phosphorus contained in the microbial biomass (pool sizes) is $9.5\,\mathrm{g\,m^{-2}}$ and $1.1\,\mathrm{g\,m^{-2}}$ respectively. Assuming the turnover time (section 5.3) of the microbial biomass to be $2.5\,\mathrm{yr}$ (as measured for biomass carbon) then the annual fluxes of these elements through the microbial biomass can be estimated to be $3.8\,g\,N\,m^{-2}$ and $0.5\,g\,P\,m^{-2}$.

The annual removal of nutrients at harvest from the unmanured plot at Broadbalk has been estimated to be $2.4\,g\,N\,m^{-2}$ and $0.5\,g\,P\,m^{-2}$, indicating that the flux of nutrients through the microbial biomass makes an important contribution to nutrient supply to the crop. The origin of this input of nitrogen could be free-living nitrogen-fixing organisms (sections 2.4.1 and 3.2.1) and rainfall (section 4.3.3); the phosphorus is probably derived from mineralisation or solubilisation of otherwise unavailable forms of phosphorus.

1.4.1 Food webs

Studies on protozoa and nematodes in microcosms indicate that the interaction of these organisms with bacteria may be important in enhancing mineralisation of nutrients from organic matter, and in some cases increasing uptake of nutrients by plants. The grazing of bacteria by protozoa will be greatest where bacterial growth is occurring most rapidly, for example, in the rhizosphere (section 3.1) and on freshly added organic substrates such as manure. In the rhizosphere there is likely to be adequate carbon for growth but a limited supply of nitrogen; grazing protozoa may release, as ammonia, approximately one third of the nitrogen immobilised in bacterial biomass close to the roots. It has been shown in microcosms that the presence of protozoa increases mineralisation and uptake of nitrogen by plants.

There is some evidence of interactions between microfauna and bacteria in field soils. After a dry soil is re-wetted during rainfall the numbers of bacteria increase, and this is usually followed by a rapid decline in numbers. Protozoa have been implicated in this decline in bacteria, which is unlikely to be due to either autolysis of cells (conditions should be favourable for bacteria), or a major bacteriophage (virus) attack. Furthermore, generation times for nematodes (also potential predators) are too long to allow a rapid increase in grazing pressure.

In the humus layer of a Swedish forest soil (pH 3.5–4.0), where most of the feeder roots were located, an increase in bacteria was observed two days after rainfall, followed by an increase in the population of naked amoebae after five days. Sixty per cent of the subsequent decrease in the bacterial population could be accounted for by the increase in protozoa. Also, data from Sweden for a 120-year-old pine forest indicated that the fauna $(1–7\,g\,m^{-2})$, although only a small proportion of the total biomass compared to bacteria $(39\,g\,m^{-2})$ and fungi $(120\,g\,m^{-2})$, consumed 30–60% of

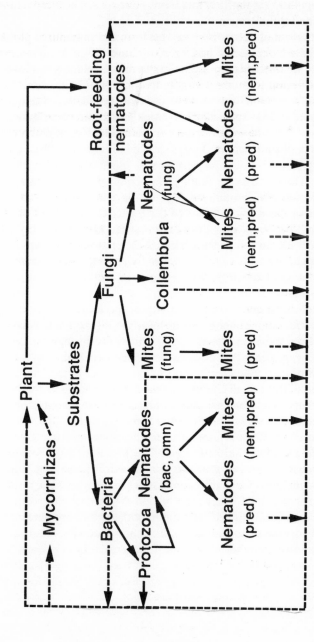

Figure 1.7 Flux pathways for nitrogen in a short grass prairie in the USA (*BAC* bacterivorous, *FUNG* fungivorous, *OMN* omnivorous, *PRED* predatory, *NEM* nematophagous, *solid lines* organic nitrogen, *broken lines* inorganic nitrogen). Fluxes from all organisms to the substrate pool (death) have been omitted. (Based on H.W. Hunt *et al.* (1987) *Biology and Fertility of Soils* **3**, 57–68.)

the microbial population in the litter and humus horizons, which accounted for 10–49% of the total nitrogen mineralised ($2.8 \, g \, m^{-2} \, yr^{-1}$).

More detailed information on the flow of energy and nutrients between functional groups of organisms has been obtained from studies of a shortgrass prairie soil in Colorado, USA. The dominant plant is *Bouteloua gracilis*, and the biomass of the main functional groups are given in Table 1.8. Using a combination of field data on population sizes and assumptions about carbon content, C:N ratios, death rates and assimilation efficiencies of the different groups, an annual budget for nitrogen was calculated based on the nutrient pathways shown in Figure 1.7. These data show that the greatest amount of nitrogen is mineralised by bacteria ($4.5 \, g \, N \, m^{-2} \, yr^{-1}$) followed by the fauna ($2.9 \, g \, N \, m^{-2} \, yr^{-1}$). Amoebae and bacterivorous nematodes account for 83% of the mineralisation by the fauna. Very little of the nitrogen flows to the higher trophic levels of the mites and collembola. Values for direct mineralisation by fauna do not reflect their importance in this process because they return a large amount of nitrogen as labile substrates which are rapidly mineralised by bacteria. The role of the fauna is therefore to accelerate the recycling of nutrients within the ecosystem.

It has been suggested that a greater variety of species in a community leads to reduced fluctuations in population densities and the greater variety of channels in a food web stabilises the flow of nutrients through that web. Figure 1.7, which represents the food web in a grassland soil, shows that compartmentalisation of food channels occurs at the base of the food web where the bacterial and fungal channels are quite separate, and the only common consumers, predatory nematodes and mites, are several trophic links away. There is also temporal compartmentalisation due to the seasonal pattern of substrate availability, and habitat compartmentalisation; the bacteria-based channels, involving bacteria, protozoa and nematodes all require a film of moisture for growth, whereas the fungi-based channels which involve mainly fungi, mites and collembola, do not require a continuous film of water (section 4.6). More data are required from a range of ecosystems to test fully the hypothesis that an increase in species diversity increases the stability of an ecosystem, when associated with an increase in the number of channels of energy flow.

CHAPTER TWO
SOIL BIOCHEMISTRY

In Chapter 1 the soil population was linked with the cycling of nutrients, in particular the mineralisation of elements such as carbon and nitrogen from organic compounds. This chapter examines the nutrition of the soil population, some of the biochemical reactions involved, and the implications of these reactions for the cycling of nutrients in soil.

2.1 Substrate supply

The importance of photosynthesis in providing a supply of energy-rich organic materials to the soil was considered in section 1.3.1. Plants are the ultimate source of energy for most of the soil population (section 2.3.1) and the amount of plant material entering the soil will play a large part in determining soil microbial activity.

2.1.1 *Sources of substrates*

Plant material can enter the soil by a variety of routes (Figure 2.1). In natural ecosystems shoot material (*litter*) forms the major input of organic material. If it is assumed that the ecosystems in Table 1.4 are at steady state (section 5.3) then net primary production (estimated from shoot data) gives the potential input of organic matter into the soils. Some of this enters the soil indirectly, following grazing by animals, as excreta or corpses. For example, in grazed hill and upland areas of the UK up to 80% of the organic nitrogen fixed by white clover (*Trifolium repens*) and returned to the soil passes through the grazing sheep. The relatively resistant soil organic matter will also provide substrates, although the distinction between resistant plant and animal material and humic material is not clear (section 2.2.2).

Where annual crops such as wheat (*Triticum aestivum*) or beans

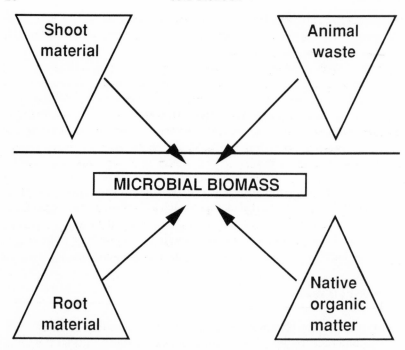

Figure 2.1 Sources of organic substrates for soil micro-organisms.

(*Phaseolus vulgaris*) are grown most of the shoot material is removed from the land at harvest, particularly in tropical regions, therefore the input of organic material is largely restricted to that derived from roots. This includes material lost from the roots while the plants are growing and from the dead roots remaining at harvest. The practical difficulties of studying roots growing in soil means that their contribution to the input of organic material into soils in the field is uncertain.

The amount of carbon lost from roots has been measured under laboratory conditions for cereals growing in solution and sand culture, and in soil using ^{14}C tracer techniques. Estimates for total carbon lost (organic carbon plus carbon dioxide) from plants supplied with ^{14}C-labelled carbon dioxide range from 14–40% of total net carbon fixed by photosynthesis (i.e. allowing for shoot respiration). Some of the carbon loss measured by this technique may be in whole roots not removed from soil, which cannot be considered as rhizosphere (section 3.1). Loss of material is generally greater in non-sterile than sterile systems and for roots growing in solid media rather than in solution culture. Losses are also increased by anaerobiosis, moisture stress, low temperatures, removal of shoots and herbicides. Most work has been on seedling plants, however, the nature and amount of material may change as the plant ages.

There is little information on losses of root material from trees. The formation of an ectomycorrhizal sheath around the roots of some trees such as beech (*Fagus sylvatica*) may represent an adaptation by the fungus to maximise its utilisation of tree root exudates. If this is so then the mycorrhizal fungal biomass associated with trees is an indication of the loss of organic material from tree roots. Estimates of fungal biomass carbon production as great as $1542 \, \text{g m}^{-2} \, \text{yr}^{-1}$ have been reported (Table 3.5).

Field measurements of losses of organic material from roots are more difficult to make, particularly when carbon fluxes must be partitioned into those due to root respiration, uptake of carbon dioxide by autotrophic organisms (a minor component) and respiration by heterotrophic micro-organisms utilising either root-derived material or existing soil organic matter (Figure 2.2). The limited data available suggest that the major flux of

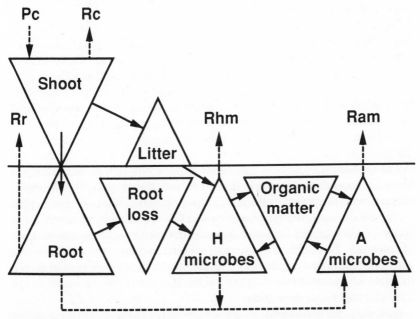

Figure 2.2 The transformations of carbon in an arable soil (*RC* respiration by the canopy, *PC* photosynthesis by the canopy, *RR* root respiration, *RHM* respiration by the heterotrophic (*H*) micro-organisms, *RAM* respiration by the autotrophic (*A*) micro-organisms, *solid lines* fluxes of organic C, *broken lines* fluxes of CO_2). (From M. Wood (1987) *Plant and soil* **97**, 303–314.)

carbon dioxide in the rhizosphere is due to microbial decomposition of root material. It has been estimated that 370–960 g m^{-2} root-derived material is lost from a crop of barley (*Hordeum vulgare*) growing in the UK.

2.1.2 *Amounts of substrates*

Bacteria require a supply of substrates for *maintenance* of their cells (for example repair of DNA) as well as for *growth* (increase in cell size and number), and the concepts developed from pure culture studies may be applied to bacteria in soil. The substrate requirements for a bacterial population can be expressed as:

$$dx/dt + ax = y \, (ds/dt)$$

where dx/dt is the growth rate, x is the biomass, a is the specific maintenance rate, y is the yield of bacteria per unit substrate, and ds/dt is the rate of input of substrate. There is little information on maintenance rates or yield coefficients for bacteria in soil, although a yield coefficient of 0.35 is often assumed (section 2.2.1). If a maintenance rate of 350 g g^{-1} yr^{-1} (0.04 g g^{-1} h^{-1}) is assumed, then the bacterial population in the UK forest soil described in Table 1.7 would require 1300 g m^{-2} yr^{-1} for maintenance alone. The estimated input of substrate as above-ground litter (545 g m^{-2} yr^{-1}) and root-derived material (162 g m^{-2} yr^{-1}) of 707 g m^{-2} yr^{-1} is therefore apparently insufficient to support this bacterial population. Similarly, for a Canadian grassland soil the estimated maintenance requirement for a bacterial population of 55 g m^{-2} is 19 270 g m^{-2} yr^{-1} compared to an estimated input of substrate of 500 g m^{-2} yr^{-1}.

Calculations such as these lend support to the view that growth of the soil bacterial population is severely limited by substrate supply. When similar calculations are applied to the soil fungal population (measured as hyphal length by the agar film technique) in the UK woodland soil described in Table 1.7, the estimated maintenance requirements exceed the annual input of substrate. This reflects our lack of understanding of the nutrition and activity of the fungal population. The data for the US grassland soil shown in Table 1.8 assumed that only 10% of the fungal hyphae measured by the agar film technique were active.

Soil may be considered as an *oligotrophic* environment, a concept developed for nutrient-poor aquatic ecosystems. The bacteria isolated from oligotrophic environments show relatively high growth rates at low concentrations of substrates and low maximum growth rates, and they can be isolated on low nutrient media. If it is assumed that substrates are evenly distributed in soil, and that all soil bacteria are equal in biomass, ATP content, metabolic activity and ability to compete for limited nutrients, then soil does appear to be an oligotrophic environment. Many soil bacteria appear unable to grow on high nutrient media; estimates of bacterial populations in soil using plate counts are considerably lower than those obtained by direct microscopy, but are increased by using low nutrient media. A low nutrient soil extract medium is recommended for the isolation of *Arthrobacter sp.* from soil. The concept of an oligotrophic soil bacterial population is similar in many respects to that of the *autochthonous* population (section 1.3.2.2).

Competition between organisms for nutrients in soil is likely to lead to specialisation in terms of rate of growth and substrate utilisation. Rather than the whole of the soil bacterial population growing slowly and continuously (the *oligotrophic* or *autochthonous* population), part of the population may have periods of dormancy interspersed with periods of

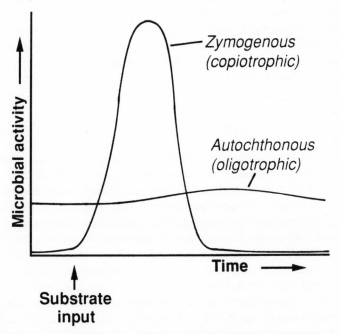

Figure 2.3 Schematic diagram of the response of two different components of the soil microbial population to an input of substrate.

relatively rapid growth (the *copiotrophic* or *zymogenous* population). This may lead to a succession of organisms in response to an input of organic material in which the zymogenous population responds rapidly to the readily available substrates, followed by the slower-growing autochthonous population utilising less readily available substrates (Figure 2.3). A similar scheme has been proposed for fungi, in which the rapidly growing sugar fungi are the primary colonisers, utilising simple compounds, followed by the slower-growing cellulolytic and ligninolytic fungi. The decomposition products of the secondary colonisers may provide further substrates for the initial colonisers. This, together with seasonal fluctuations in supply of litter and root material produces temporal variation in substrate supply.

There is also considerable spatial variability in substrate supply in soil. Litter and animal excreta are not uniformly distributed on the soil surface, and roots provide localised zones of high nutrient concentration. At a smaller scale bacterial cells provide substrate for organisms such as the predatory bacterium *Bdellovibrio sp.*, and the cell walls of dead fungal hyphae are attacked by streptomycetes. Exudates from fungal spores may also provide high local concentrations of substrates. Nutrients may accumulate at interfaces such as the negatively-charged surfaces of clays and organic matter (section 4.9). It is therefore an over-simplification to consider the soil as a uniformly oligotrophic environment.

2.2　Substrate quality

The input of organic material into soil from living and dead organisms provides substrates for growth of organisms (*detritivores*), primarily microorganisms, and subsequently for the mesofauna, macrofauna and flora. However, the composition of this organic material reflects that of the organisms from which it is derived. It therefore varies greatly in its content of elements such as nitrogen and phosphorus, and also in the types of molecules it contains. For example, soluble organic material lost from roots contains a wide range of compounds including sugars such as glucose and arabinose, amino acids such as glutamic acid and leucine, organic acids such as oxalic and propionic acids, nucleotides such as adenine and guanine, and enzymes including invertase and protease. Simple soluble molecules such as glutamic acid are readily assimilated by a wide range of organisms, whereas the ability to metabolise the complex and insoluble molecule lignin is restricted to a few fungi. The quality of substrates can therefore be considered in terms of elemental composition and availability.

2.2.1 Elemental composition

Table 2.1 shows the major elemental content of a range of organic materials. Because of the major technical difficulties in directly extracting micro-organisms from soil the data for the bacterium and fungus are obtained from laboratory cultures of these organisms. *Escherichia coli* is not normally found in soil, but data are not available for common soil bacteria. All four materials are dominated by carbon which comprises nearly half the dry weight. The other major elements present are oxygen, nitrogen, hydrogen, phosphorus and sulphur. Transformations of these elements, which make up about 95% of the cell dry weight, will be linked with mineralisation of organic compounds. This is illustrated by the link between carbon and nitrogen mineralisation.

A bacterial cell in soil can obtain nutrients from either organic sources such as amino acids, or from inorganic sources such as ammonium ions, but it must assimilate elements in a ratio approximately equal to that required for the elemental composition of its own cell. For example, carbon and nitrogen must be assimilated in a ratio of approximately 5:1 (Table 2.1). If a bacterium is to utilise all of the carbon in a substrate such as maize shoots, with a C:N ratio of 31.4, then it must obtain extra nitrogen from inorganic sources to satisfy its own C:N ratio. The uptake and assimilation of inorganic nutrients by bacteria during the decomposition of organic materials in soil is termed *immobilisation*.

During decomposition of organic material by micro-organisms some of the carbon in the substrate is released as carbon dioxide by respiration, therefore as less of the carbon is assimilated, less nitrogen is required to meet the required C:N ratio. The proportion of carbon assimilated from a substrate (i.e. yield) varies but a value of 0.35 has often been used to estimate bacterial growth rates in soil (section 2.1.2). The new bacterial biomass

Table 2.1 Elemental composition (% dry weight) of some organic materials. After Jenkinson and Ladd (1981).

Material	Carbon	Nitrogen	Phosphorus	Sulphur	C:N ratio
Bacterium (*Escherichia coli*)	50	15	3.2	1.1	3.3
Fungus (*Penicillium chrysogenum*)	44	3.4	0.6	0.4	12.9
Maize (*Zea mays*) shoots	44	1.4	0.2	0.2	31.4
Farmyard manure (cattle on straw)	37	2.8	0.5	0.7	13.2

eventually serves as a substrate with a relatively low C:N ratio for other functional groups such as protozoa and nematodes which have C:N ratios of < 10:1. Release of some of the carbon during respiration by the assimilating organism will now lead to an excess of nitrogen relative to carbon in the substrate, and part of the nitrogen is released into the soil in the form of ammonium. The release of inorganic nutrients during the decomposition of organic materials is termed *mineralisation*.

Carbon released as carbon dioxide during the initial decomposition stage will also reduce the effective C:N ratio of the substrate; if the substrate has a high initial C:N ratio, as in the case of maize shoots, this will reduce the effective C:N ratio of the substrate to a value which still remains higher than that of the assimilating organisms, resulting in immobilisation. The inefficient use of carbon by some organisms which release a large amount of carbon dioxide per unit substrate (i.e. have a low cell yield) will reduce further the effective C:N ratio of the substrate. The fixation of atmospheric nitrogen gas by some bacteria (section 2.4.1) provides an alternative strategy for utilizing substrates with a high C:N ratio.

Mineralisation is brought about by a succession of organisms in soil (section 1.4.1). The role of higher trophic levels in determining mineralisation is illustrated in Figure 2.4. The relatively low efficiency of carbon assimilation by nematodes (approximately 10%) together with their feeding on microbial biomass with a relatively low C:N ratio suggests an important and perhaps overlooked role for this group of organisms in mineralisation of nitrogen (section 1.4.1).

Immobilisation can have important consequences for plants. For example, the incorporation of straw with a high C:N ratio into soil before planting a crop may lead to immobilisation of the mineral nitrogen which would otherwise be available to support plant growth, and the crop may become nitrogen deficient. However, immobilisation may be beneficial. For example, in temperate regions the cultivation of land in the autumn causes a flush of mineralisation from disturbed organic matter and the nitrate produced by oxidation of the ammonium thus formed may be leached from the soil and eventually enter water supplies (section 6.2.1). Immobilisation of this mineral nitrogen into microbial biomass reduces this risk and allows a steady re-mineralisation of the nitrogen from the microbial biomass for crop growth during the following season. These two examples illustrate the fine balance between mineralisation and immobilisation which often occurs in soil. Both processes may operate concurrently in a soil (this is referred to as *mineralisation-immobilisation turnover* or MIT) and the net result determines the availability of a particular nutrient to plants.

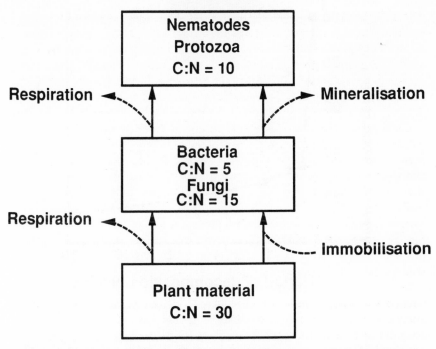

Figure 2.4 The role of different groups of soil organisms in the mineralisation and immobilisation of soil nitrogen.

Fungi generally have a higher C:N ratio than bacteria (Table 2.1) and may therefore immobilise less nitrogen per unit substrate assimilated than bacteria. However, such differences may be offset by the fungi having a higher efficiency of carbon assimilation (less carbon lost as carbon dioxide). Animal manures with a lower C:N ratio are less likely to cause immobilisation than plant materials (Table 2.1). The residue of a legume such as medic (*Medicago littoralis*), with a C:N ratio of 15:1, is likely to mineralise nitrogen from the start of decomposition, differing in this respect from cereal or grass residues, which normally immobilise nitrogen during the early stages of decomposition.

2.2.2 *Availability of substrates*

In the examples used in section 2.2.1 to illustrate the effect of the elemental composition in determining the fate of a substrate, it was assumed that all of the carbon and nitrogen in the substrate was immediately available to the

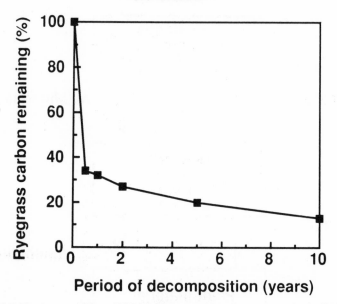

Figure 2.5 The decomposition of ^{14}C-labelled ryegrass (*Lolium perenne*) in the field. (After D.S. Jenkinson (1981).)

bacteria. However, studies using tracers such as ^{14}C and ^{15}N to follow the decomposition of organic materials in soil indicate otherwise.

Figure 2.5 shows the decomposition of ryegrass (*Lolium perenne*) labelled with [14]C. Decomposition appears to follow two distinct phases: a relatively short and rapid initial phase, followed by a longer, slower phase. During the first year about two-thirds of the carbon is lost from the plant material. Some of the breakdown products from this material are converted into more resistant humic material and this new material, together with more resistant plant material is decomposed during the second phase. Clearly a fraction of the carbon in the substrate remains unavailable to the microbial biomass for a considerable time. Decomposition of ^{14}C, ^{15}N-labelled medic (*Medicago littoralis*) in South Australian soils followed a similar pattern. More than 50% of the medic ^{14}C had disappeared after four weeks and only 15–20% of ^{14}C and 45–50% of ^{15}N remained as organic residues after four years.

Table 2.2 shows the sizes of the estimated substrate pools for a Canadian pasture soil. The native organic matter has been partitioned according to anticipated availability to the microbial population. It has been estimated that 50–60% of soil organic matter is protected from decomposition due to

Table 2.2 Major sources and estimated amounts of organic substrates in a Canadian pasture soil. After E.A. Paul and R.P. Voroney (1980), in *Contemporary Microbial Ecology*, ed. D.C. Ellwood *et al.*, Academic Press, New York, 215–237.

Source	Amount $(g\,C\,m^{-2})$
Litter	100
Root	1030
Microbial biomass	125
Organic matter	
Decomposable (not protected)	440
Decomposable (protected)	4410
Recalcitrant (not protected)	1700
Recalcitrant (protected)	1700

its association with soil solids. For example, amino acids adsorbed to clay particles are degraded much more slowly than free amino acids. The proportion of those parts of perennial plants that becomes available to the microbial biomass annually is not clear, although a decomposition rate of $0.08\,g\,g^{-1}\,day^{-1}$ has been used in a simulation model of carbon turnover in this soil.

In mature temperate and tropical forest soils 92–99% of the above-ground biomass is in the form of *wood*, and a high proportion of the root system of trees is also likely to be wood. Although the input of wood as a substrate for soil organisms is often underestimated, woody tissues such as branches are conserved by plants, whereas photosynthetic tissues such as leaves are not. Wood is also a low quality resource due to extensive lignification, low content of soluble sugars and minerals, an often high content of allelopathic compounds (section 3.1.4), protection by bark, and the small external area to volume ratio associated with the often massive size of the substrate units (an oak tree may be $50\,m^3$).

Although many basidiomycete fungi possess the ability to decompose wood completely, under natural conditions decomposition occurs as a result of interactions between animals and fungi. The initial 60% of weight loss of wood in a temperate deciduous forest, which occurs mainly as death and decay in the canopy and branch fall, is due to fungi. The final 40% decomposition is due to the cooperative action of wood-boring animals and fungi. The fungal contribution may involve enzymatic softening of the wood, destruction of allelopathic compounds, and the production of attractants. Fungi may also improve the nutritional quality of the wood by reducing the C:N ratio. For example, in order to gain 10 μg of nitrogen an animal would need to consume 5.3 mg of undecayed wood, but only 1.9 mg of wood which had already lost 60% of its weight due to fungal colonisation. A considerable fraction of the nutrients in wood consumed by animals may be in the form of fungal mycelium. Finally, the penetration of bark by animals such as wood-borers and woodpeckers may allow colonisation by a secondary fungal population.

Termites are involved in decomposition of wood in tropical soils, and may attack wood before it falls to the ground. Conditioning of substrates by microbes prior to consumption by these animals is also important (section 5.3.2).

2.3 Nutrition of soil bacteria

Bacteria are the most biochemically diverse group of organisms in soil and many of the strategies they adopt to obtain their nutrients have important consequences for nutrient cycling in soil. Figure 2.6 shows the main nutrients required by bacteria. 80–90% of total cell weight is water. Micronutrients include iron, copper and molybdenum and some organisms

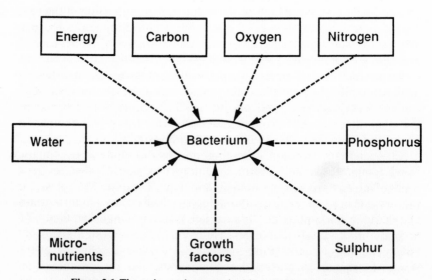

Figure 2.6 The main nutrients required by soil micro-organisms.

require pre-formed compounds (cofactors) such as vitamins. Energy is required for life, but not all bacteria require atmospheric oxygen.

2.3.1 Energy and carbon

Bacteria employ three methods for obtaining energy, these are: fermentation, respiration, and photosynthesis. In all these processes the harvested energy is stored in the form of adenosine triphosphate (ATP) and used for biosynthesis. ATP is generated as electrons are transported by *reduction-oxidation reactions* via a series of compounds to a *terminal electron acceptor*. The flow of electrons across the cytoplasmic membrane is coupled to a net translocation of protons from the cytoplasm to the external aqueous phase. The resulting *proton electrochemical gradient* (also known as the *proton motive force*) drives protons through the ATP synthase enzyme associated with the membrane thereby generating ATP. The proton electrochemical gradient also drives active transport to and from the cell, the movement of flagella (section 4.7) and reverse (endergonic) electron transport (section 2.4.2).

Fermentation (substrate level phosphorylation) is the simplest method of ATP generation in which organic compounds serve as electron donors (are oxidised) and electron acceptors (are reduced). However, the average oxidation state of the products is the same as that of the substrates. It can proceed in the absence of oxygen, but the process is relatively inefficient and yields only $2\,mol$ ATP mol^{-1} glucose. Carbohydrates are the main substrates yielding products such as acetic acid (section 3.2.2), but bacteria can also ferment organic acids and amino acids.

Respiration (oxidative phosphorylation) is a more complex and efficient method of generating ATP in which organic or inorganic compounds serve as electron donors (are oxidised) and inorganic compounds serve as the terminal electron acceptors (are reduced). The use of NH_4^+ and NO_2^- as electron donors forms the basis of nitrification. Under aerobic conditions oxygen serves as the terminal electron acceptor (*aerobic respiration*) and under anaerobic conditions NO_3^- or SO_4^{2-} and other substances may serve in place of oxygen (*anaerobic respiration*). The use of NO_3^- or SO_4^{2-} as terminal electron acceptors forms the biochemical basis of the important soil processes of denitrification and sulphate reduction respectively (section 2.4). Some organisms can grow either in the absence or presence of oxygen (*facultative anaerobes*) and some can only grow in the absence of oxygen (*obligate anaerobes*). Aerobic respiration yields $38\,mol$ ATP mol^{-1} glucose.

| Energy source | Carbon source | |
	CO_2 *autotroph*	Organic C *heterotroph*
Light *Photo-*	*Photoautotroph* **Algae Plants**	*Photoheterotroph* **Green and purple bacteria**
Chemical *Chemo-*	*Chemoautotroph* **Nitrifying and S oxidising bacteria**	*Chemoheterotroph* **Most bacteria Fungi Protozoa**

Figure 2.7 Groupings of soil organisms according to their source of carbon and energy.

In *photosynthesis* (photophosphorylation), light energy is absorbed by a photosynthetic pigment system, including chlorophyll, and electrons are transported to water, the terminal acceptor, which leads to oxygen production. Organisms which obtain energy by this method are termed *phototrophs*, to distinguish them from those organisms which obtain energy from chemical sources by respiration and fermentation (*chemotrophs*). Organisms may obtain carbon from either organic compounds (*heterotrophs*) or from carbon dioxide (*autotrophs*). Fixation of carbon dioxide by micro-organisms involves the same enzymes, for example ribulose bisphosphate (RuBP) carboxylase, as in plants. Organisms may therefore be grouped according to the sources from which they obtain their energy and carbon (Figure 2.7). *Photoautotrophs*, for example maize (*Zea mays*), obtain energy from light and carbon from carbon dioxide, and *photoheterotrophs*, for example the green bacterium *Chlorobium sp.*, use light energy and organic carbon. *Chemoautotrophs* (also termed *chemolithotrophs*) obtain energy from chemical reactions and carbon from carbon dioxide, for example the bacterium *Nitrosomonas europea*, obtains energy from the oxidation of NH_4^+ to NO_2^-, and *Nitrobacter sp.* from the oxidation of NO_2^- to NO_3^-; these reactions constitute nitrification, an important component of the nitrogen cycle in soils (section 2.4.2). *Chemoheterotrophs*, for example the fungus *Trichoderma harzianum*, obtain both energy and carbon from organic compounds.

2.3.2 Nitrogen

NH_4^+ is the preferred source of nitrogen for bacteria although organisms containing nitrate reductase (a molybdenum-containing enzyme which reduces NO_3^- to NH_4^+) can also assimilate NO_3^-. NO_3^- assimilation is a highly regulated process which normally proceeds slowly at the rate at which NH_4^+ is required for growth. NO_2^- (an intermediate) rarely accumulates; this contrasts with dissimilatory nitrate reduction (section 2.4.3) which is more rapid, and which can lead to an accumulation of NO_2^-. Plants can take up either NH_4^+ or NO_3^-, but where the supply of NO_3^- is limited, for example in acid soils (section 4.3.2), then organisms may utilise mainly NH_4^+. Heterotrophic organisms may obtain nitrogen from organic compounds such as peptides and amino acids. A single compound such as an amino acid can provide a micro-organism with energy, carbon and nitrogen. A few genera of bacteria are able to utilise atmospheric nitrogen gas (nitrogen fixation, section 2.4.1).

Within the microbial cell all sources of nitrogen are converted into NH_4^+, and then into glutamate and glutamine, which are key intermediates in subsequent biosynthesis of nitrogen compounds. It is not clear how ions such as NO_3^- and SO_4^{2-} are taken into a cell which has a membrane potential which is negative inside the cell.

2.4 Selected biochemical processes

All biochemical processes rely upon catalysts to accelerate the rate of reaction. Biological catalysts are termed *enzymes*. They are proteins, and like inorganic catalysts they remain unchanged after completion of the reaction. The sequence of amino acids which make up a such a protein is determined by the genetic information carried in the cell in the form of *deoxyribonucleic acid* (DNA). A unique sequence of bases in the DNA (a *gene*) codes for a particular amino acid. Some genes influence the *expression* of other genes, thereby controlling enzyme synthesis or activity. Some of the biochemical processes which have been mentioned in relation to the nutrition of soil bacteria are now considered in more detail. In some cases details of the *genetics* of the processes are included.

Although it is possible to present nutrient transformations as simple cycles (Figure 1.6 and Figure 1.7), detailed investigations of the biochemical pathways for the processes and the ecology of the organisms (section 4.8) have revealed a more complex nature to the transformation of elements in soil. Processes have been selected here which are particularly important in the transformation of nitrogen and sulphur in soil.

2.4.1   Nitrogen fixation

Nitrogen fixation is a reaction used by some bacteria to obtain nitrogen. It can be represented by the following equation:

$$N_2 + 6H^+ + 6e^- \rightarrow 2NH_3$$

The atmosphere contains a large pool of nitrogen in the form of nitrogen gas (79% of the total gases), but the ability to reduce nitrogen to NH_3 by nitrogen fixation is restricted to certain prokaryotic organisms. These include the free-living aerobic bacterium *Azotobacter sp.*, the obligate anaerobe *Clostridium sp.*, the photosynthetic blue-green bacterium *Anabaena sp.*, and the symbiotic bacteria *Rhizobium sp.* and *Frankia sp.*

The N_2 molecule is extremely stable; the industrial production of NH_3 for fertiliser from N_2 by the Haber–Bosch process requires a temperature of 400°C and a pressure of 20–35 MPa (200–350 atmospheres) together with a finely divided iron catalyst to break the triple covalent bond between the nitrogen atoms. Bacteria are able to carry out this reaction at normal temperature and pressure using an enzyme. This enzyme, termed *nitrogenase*, is similar in all nitrogen-fixing organisms. It consists of two proteins, a large one containing iron and molybdenum, and a small one containing iron both of which are destroyed by oxygen.

Nitrogen-fixing organisms growing in an aerobic environment have evolved a variety of mechanisms to protect nitrogenase from oxygen, for example the production of leghaemoglobin in legume root nodules (section 3.4.2). The free-living bacterium *Azotobacter sp.* maintains a low intracellular oxygen concentration by having a high rate of respiration which can be uncoupled from ATP generation to prevent excess ATP production. In photosynthetic nitrogen-fixing organisms this problem is exacerbated by the production of oxygen during photosynthesis. Cyanobacteria (blue-green bacteria) have specialized cells (*heterocysts*) in which nitrogen fixation occurs but which do not photosynthesise. Fixed carbon and nitrogen are transported to and from heterocyst cells.

The biological reduction of nitrogen also requires a large amount of energy; 12–15 ATP molecules are required for the reduction of one molecule of N_2. ATP in the form of MgATP binds to the iron-containing protein and stimulates transfer of electrons obtained from reduced ferredoxin and flavodoxin to iron atoms in the molybdenum–iron protein.

At least one ATP is hydrolysed for each electron transferred. N_2 binds to the active site of the molybdenum–iron protein and is reduced to a series of enzyme-bound dinitrogen hydride intermediates (the exact identity of these is unknown) and eventually to ammonia, which is assimilated as glutamate and glutamine.

Nitrogenase also reduces other triply-bonded molecules such as acetylene to ethylene. This forms the basis of the *acetylene reduction assay* which has been used extensively to measure rates of nitrogen fixation. Problems associated with the use of this method to measure rates of nitrogen fixation in the field include endogenous ethylene production by soil microorganisms, inhibition of hydrogen evolution from nodules and non-linear production of ethylene with time. Such measurements provide only short term estimates of nitrogenase activity and large errors are associated with extrapolation of data to seasonal estimates. The only direct measurement of nitrogen fixation is given by $^{15}N_2$ reduction in which a legume or other nitrogen-fixing system is incubated in an atmosphere enriched with the stable isotope ^{15}N. The amount of isotope fixed is measured by mass spectrometry (section 3.2.1). This method is expensive and not suitable for use in the field, but it is essential for calibration of other techniques.

Nitrogenase can also reduce H^+, which leads to hydrogen evolution (section 3.4.3). It is also known that some forms of nitrogenase exist which contain vanadium rather than molybdenum, although the detailed chemistry of this enzyme is not known.

The genetics of nitrogen fixation have been intensively studied. The production and regulation of nitrogenase is controlled by a closely linked series of 17 *nif* genes. *nif* D and *nif* K control synthesis of the molybdenum–iron proteins and *nif* B is involved in the construction of iron-molybdenum cofactors. *nif* A is a regulatory gene which may be activated by plant compounds such as flavonoids, similar to the *nod* D gene (section 3.4.2). There is also a series of *Fix* genes which are necessary for nitrogen fixation, but their role is unclear.

2.4.2 *Nitrification*

Nitrification involves two reactions used by only a few genera of autotrophic bacteria to generate energy. The energy yield for the reaction is low, resulting in very slow growth rates. It can be represented by the following overall equation:

$$NH_4^+ \rightarrow NO_2^- \rightarrow NO_3^-$$

Ammonium oxidation is restricted to five genera of bacteria, for example, *Nitrosomonas sp*. The energy yield is only $272 \, kJ \, mole^{-1}$ of NH_4^+ oxidised compared to the oxidation of glucose which yields $2872 \, kJ \, mole^{-1}$. Hydroxylamine (NH_2OH) is an intermediate in the reaction. The reaction generates acidity as protons are released:

$$NH_4^+ + \tfrac{3}{2}O_2 \rightarrow NO_2^- + H_2O + 2H^+$$

NH_4^+ contains nitrogen in its most reduced state, but it is not readily oxidised. The direct oxidation of NH_4^+ to NH_2OH by oxygen is endergonic, and therefore requires a special type of enzyme or chemically reactive species. Once the first $N-O$ bond is formed the subsequent oxidation is more favourable which allows NH_4^+ oxidation to NO_2^- to be used as an energy source.

In autotrophic nitrifying bacteria hydroxylamine is formed by *ammonium monooxygenase* (a copper-containing enzyme):

$$NH_4^+ + O_2 + H^+ + 2e^- \rightarrow NH_2OH + H_2O$$

The reaction is a reduction and relies upon electrons produced by the subsequent oxidation of NH_2OH:

$$NH_2OH + H_2O \rightarrow NO_2^- + 5H^+ + 4e^-$$

A small proportion of the electrons transported across the membrane by this reaction are transported back (*reverse electron transport*) to generate reductants to allow the coversion of NH_4^+ to NH_2OH. This is an unusual reaction because the cell relies upon the monooxygenase to produce hydroxylamine which is the electron donor for the reaction.

Nitrite oxidation, which is even less energetically favourable (71 $kJ \, mole^{-1}$ of NO_2^-), is restricted to one genus of bacteria, *Nitrobacter*:

$$NO_2^- + \tfrac{1}{2}O_2 \rightarrow NO_3^-$$

Reverse electron transport of approximately one-fiftieth of the electrons produced from NO_2^- is used to generate reductant (NADH) which may be used, for example, in assimilation of carbon from carbon dioxide.

There is evidence that some heterotrophic organisms, both bacteria and fungi, also oxidise NH_4^+ and certain organic nitrogen compounds to produce NO_2^- and NO_3^-, although the biochemical basis of this reaction is not known. It may be that the free radicals produced during the enzymatic degradation of lignin (section 2.5.4) also oxidise NH_4^+. In the laboratory, *methanotrophs* (bacteria able to grow on methane) are able to oxidise NH_4^+ using *methane monooxygenase*, but the extent to which this occurs in the field is not clear. Heterotrophic nitrifiers such as the fungus *Absidia sp.* may be important in acid forest soils (section 4.3.2).

2.4.3 *Denitrification*

Denitrification involves a series of reactions used by a wide range of facultative anaerobic bacteria in the absence of oxygen in which oxidised forms of nitrogen are used as alternative electron acceptors. The overall reaction involves dissimilatory reduction of nitrogen oxides to nitrogen gases:

$$NO_3^- \rightarrow NO_2^- \rightarrow (NO) \rightarrow N_2O \rightarrow N_2$$

Specific reductase enzymes have been isolated for each of the following stages, except for nitric oxide (NO) reduction:

$$NO_3^- + 2e^- + 2H^+ \rightarrow NO_2^- + H_2O$$

$$NO_2^- + 2e^- + 2H^+ \rightarrow NO + H_2O$$

$$2NO + 2e^- + 2H^+ \rightarrow N_2O + H_2O$$

$$N_2O + 2e^- + 2H^+ \rightarrow N_2 + H_2O$$

The enzyme responsible for the first stage of denitrification, *dissimilatory nitrate reductase*, is an iron–sulphur-protein containing molybdenum (similar to nitrogenase). In *Pseudomonas aeruginosa* and *Rhizobium sp.* distinct assimilatory (section 2.3.2) and dissimilatory nitrate reductase enzymes have been identified. Dissimilatory nitrate reductase is repressed in the presence of oxygen and de-repressed in the absence of oxygen. This mechanism allows oxygen (which gives a higher molar energy yield than NO_3^-) to be used when available, also sparing NO_3^- for assimilation. However, the concentration at which de-repression occurs varies with different organisms. In soil, where the concentration of oxygen may be continuously changing, it may be an advantage for an organism to retain dissimilatory nitrate reductase activity under aerobic conditions to allow a rapid response to anaerobiosis. Under certain conditions, for example at low pH values, the reaction may stop at nitrous oxide (N_2O).

The bacteria capable of denitrification comprise some 13 genera including *Pseudomonas sp., Agrobacterium sp.* and *Bacillus sp. Bradyrhizobium japonicum* is capable of denitrification but it is not clear whether this process is linked in any way to nitrogen fixation. *Thiobacillus denitrificans* uses NO_3^- as the terminal electron acceptor during oxidation of reduced sulphur compounds such as sulphide. Under conditions of reduced oxygen concentration (but not necessarily completely anaerobic), in the presence of the appropriate bacteria and a supply of readily available organic substrate, significant quantities of nitrogen may be lost from the soil by denitrification (section 6.2.2).

Nitrous oxide may also be produced during nitrification (section 2.4.2) accounting for approximately 0.1% of the nitrogen oxidised. The content of ^{18}O and ^{15}N in nitrous oxide produced from nitrification is lower than that of nitrous oxide in the atmosphere, which suggests that the production of this gas during nitrification, probably via NO_2^-, is a minor source of atmospheric nitrous oxide compared to denitrification.

2.4.4 Sulphur oxidation

Sulphur oxidation is a reaction used by one genus of autotrophic bacteria, *Thiobacillus*, to obtain energy. It can be represented by the following overall reaction:

$$S^2 \rightarrow SO_3^{2-} \rightarrow SO_4^{2-}$$

Five species of *Thiobacillus* have been studied in detail and each one can oxidise a range of reduced sulphur compounds including elemental sulphur. All are aerobic except for *T. denitrificans*. *T. thiooxidans* and *T. ferrooxidans* have optimum pH values for growth of 2–3.5. The oxidation of elemental sulphur by *T. thiooxidans* leads to the production of acidity:

$$2S + 3O_2 + 2H_2O \rightarrow 2H_2SO_4$$

T. ferrooxidans can oxidise either reduced sulphur compounds or ferrous (Fe^{2+}) ions, also generating acidity:

$$2FeS_2 + 7O_2 + 2H_2O \rightarrow 2FeSO_4 + 2H_2SO_4$$

The oxidation of sulphur compounds involves adenosine-5'-phospho-sulphate (APS):

$$2AMP + 2SO_3^{2-} \rightarrow 2APS + 4e^-$$

$$2APS + PO_4^{3-} \rightarrow 2ADP + 2SO_4^{2-}$$

Heterotrophic organisms, both bacteria and fungi, can also oxidise sulphur compounds, but they do not obtain energy from the reaction.

2.4.5 Reduction of inorganic sulphur

SO_4^{2-} is reduced during the assimilation of SO_4^{2-} by bacteria. The assimilatory reductive pathway is closely regulated to meet the requirement for the biosynthesis of cysteine. A few genera of bacteria are also able to use SO_4^{2-} and other oxidised forms of sulphur as alternative electron acceptors for anaerobic oxidation of hydrogen and organic compounds (dissimilatory SO_4^{2-} reduction). No significant regulation of the enzymes involved in dissimilatory SO_4^{2-} reduction has been observed.

The obligate anaerobic bacterium *Desulfovibrio desulfuricans* has received most attention, although *Clostridium sp.* and *Desulfotomaculum sp.* are also important. The overall process can be represented by the following equation:

$$SO_4^{2-} \rightarrow SO_3^{2-} \rightarrow S_3O_6^{2-} \rightarrow S_2O_3^{2-} \rightarrow S^{2-}$$

The reduction of SO_4^{2-} to SO_3^{2-} involves ATP sulphurylase which catalyses the following reaction:

$$SO_4^{2-} + ATP \rightarrow APS + PP_i$$

and APS reductase:

$$APS + 2e^- \rightarrow AMP + SO_3^{2-}$$

The SO_3^{2-} is then protonated by a chemical reaction to form bisulphite (HSO_3^-) and subsequent reduction reactions are catalysed by bisulphite reductase, trithionate ($S_3O_6^{2-}$) reductase and thiosulphate ($S_2O_3^{2-}$) reductase to produce sulphide (S^{2-}).

Hydrogen sulphide (H_2S) produced under anaerobic conditions is toxic to most aerobic organisms including crop plants, and may cause precipitation of metal sulphides and hence the characteristic black colour of anaerobic muds.

2.4.6 The biochemical roles of NO_3^- and SO_4^{2-}

There are many similarities between the biochemical transformations of nitrogen and sulphur in soils. For example, nitrification and sulphur oxidation are carried out by autotrophic bacteria, and both NO_3^- and SO_4^{2-} are reduced during anaerobic respiration to form gaseous products. Figure 2.8 summarises the biochemical roles of NO_3^- and SO_4^{2-} in soils. Oxidation, assimilatory reduction (immobilisation), and dissimilatory reduction (anaerobic respiration), are reactions common to both ions.

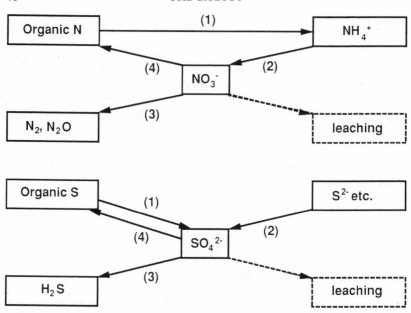

Figure 2.8 The biochemical roles of NO_3^- and SO_4^{2-} in soil (*1* mineralisation, *2* oxidation, *3* dissimilatory reduction, *4* assimilatory reduction).

2.5 Soil enzymes

The biochemical versatility of the soil bacterial population provides soil with the capability to degrade all natural compounds and most of the synthetic compounds which enter either deliberately, for example, as pesticides applied to crops, or accidentally, for example, in the industrial pollution of land. Many of the enzymes involved in these reactions are located within the organisms (*endocellular enzymes*). However, soils also possess enzyme activity which persists after the microbial population has been inhibited or killed. These are termed *extracellular* or *abiotic* enzymes.

Extracellular enzymes are mainly derived from soil micro-organisms, particularly those enzymes involved in the degradation of insoluble substrates such as proteins and carbohydrates which are too large to enter the cell and which must therefore be at least partially broken down outside the cell. Plants and animals may also produce extracellular enzymes, but there are problems in clearly distinguishing between enzymes produced by these organisms themselves and enzymes produced by associated micro-organisms in the rhizosphere and in the animal gut.

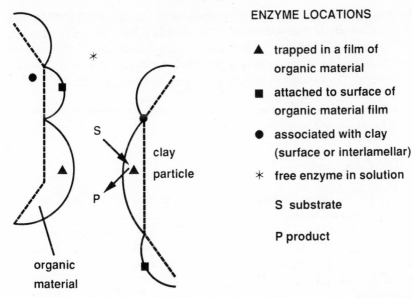

ENZYME LOCATIONS

▲ trapped in a film of
   organic material

■ attached to surface of
   organic material film

● associated with clay
   (surface or interlamellar)

✳ free enzyme in solution

S substrate

P product

Figure 2.9 A schematic diagram for the location of extracellular enzymes in soil. (After R.G. Burns (1977).)

The accumulation of extracellular enzymes in soil depends upon mechanisms for the stabilisation of these molecules, which being proteins are themselves liable to degradation by proteases. Enzymes may be adsorbed to the surface of clay minerals and between clay lamellae (section 1.2.1). Enzymes associated with organic matter are more resistant to degradation, possibly due to changes in viscosity or hydrogen bonding. The association of an enzyme with a solid surface such as a clay particle may lead to changes in substrate affinity due to unfolding or coagulation of the protein chains, or physical blocking of active sites. Alternatively, enzymes may become aligned in a way that enhances activity. A hypothetical scheme has been proposed for the stabilisation of enzymes within soil (Figure 2.9). A highly protected enzyme will be of no significance to soil enzyme activity if diffusion of substrates to the enzyme and products away from the enzyme are not possible.

Results of studies on extracellular enzyme activity in soils are highly dependent upon the methods used. In order to study such enzymes, soil endocellular activity must be inhibited without affecting the chemical and physical conditions and the extracellular enzyme activity. Antibiotics, high energy ionising radiation and a range of antiseptic and bacteriostatic agents have been used; one of the most acceptable is toluene. Furthermore, soil enzyme activity is often measured under standard but artificial conditions, for example as buffered suspensions at pH 7 with either the natural substrate or a synthetic substrate provided in excess. Such assays give an estimate of the maximum potential enzyme activity, but not the activity which actually occurs in soil.

A few of the major soil enzymes which exhibit significant extracellular activity are now considered.

2.5.1   Carbohydrases

The activities of carbohydrases such as *cellulase* are relatively easy to measure because the substrate is often insoluble, the reaction products are soluble, and neither react with soil particles. In common with many soil enzymes the activity of *invertase* (which catalyses the hydrolysis of sucrose to glucose and fructose) decreases down the soil profile, but varies considerably, and there is no correlation between activity and other soil properties.

There are many cellulolytic fungi, for example *Aspergillus sp.*, but fewer cellulolytic bacteria, although actinomycetes such as *Streptomyces sp.* produce cellulase. Cellulase is not a single enzyme but is an enzyme system, and it is normally inducible, with simple glucose-containing molecules such as cellobiose acting as inducers.

2.5.2   Esterases

Phosphatase is important, particularly in soils with low levels of available phosphorus, in the mineralisation of inorganic phosphorus from organic sources such as inositol phosphate. The measurement of phosphatase activity is complicated because inorganic phosphorus reacts with soil components and may not therefore be detected. The release of phenol from phenylphosphate is a more reliable assay for this enzyme. Different phosphatases have different optimum pH values; there are acid, neutral and alkaline phosphatases. When organisms become short of phosphorus they may increase the production of phosphatase and hence the supply of inorganic phosphorus.

Nucleases, which degrade nucleic acids, are also found in soils. However, if nucleic acids become complexed with lignin they are more resistant to enzyme attack.

Table 2.3 Urease activity (μg urea hydrolysed $g\,soil^{-1}\,h^{-1}$) in a range of soils. After J.M. Bremner and R.L. Mulvaney (1978), in *Soil Enzymes*, ed. R.G. Burns, Academic Press, 149–196.

Location	% organic C	Soil pH	Urease activity
Australia	0.16–5.88	4.8–6.7	22–416
USA	0.30–6.73	4.6–8.0	11–189
Malaysia	0.64–3.44	4.2–4.9	5–41
India	0.24–0.98	4.2–6.7	20–162

2.5.3 Proteases and amidases

Proteases and *amidases* accumulate in soil and are important in the degradation of proteins and therefore in the mineralisation of nitrogen. Urease is the soil enzyme which has been studied in most detail, mainly due to the widespread use of urea as a nitrogen fertiliser. *Urease* is stable and persistent in soil and extracellular activity may account for 50–90% of the total soil urease activity. Table 2.3 shows the urease activity measured in soils from a range of countries. The variation in activity measured for a single soil is greater than the variation between soils.

When urea fertilisers are applied to soil, ammonia is produced by enzymatic hydrolysis:

$$CO(NH_2)_2 + H_2O \rightarrow CO_2 + 3NH_3$$

Under alkaline conditions ammonia may be lost from the soil by *volatilisation*. Similarly, ammonia may volatilise from localised areas of urine deposited on the soil surface by grazing animals. It has been estimated that on average sheep void $45\,g$ urea $animal^{-1}\,day^{-1}$ and cattle void $140\,g$ urea $animal^{-1}\,day^{-1}$. The addition of urease inhibitors such as hydroquinone (section 6.2.1) to soil may reduce the losses of nitrogen by this process.

2.5.4 Oxidoreductases

Lignin is a complex polymer of aromatic nuclei, containing a basic phenyl propane unit grouped into a highly branched random structure linked by strong bonds. It is therefore extremely resistant to chemical and microbial degradation, although certain fungi, in particular white rot fungi, are ligninolytic. The enzyme responsible for lignin degradation (*ligninase*) was isolated in 1983 from the white rot fungus *Phanerochaete chrysosporium*. It contains a haem group which reacts with hydrogen peroxide to form a molecule with a high affinity for electrons which then oxidises the aromatic rings in lignin to form relatively stable free radical cations. These radicals can bring about further oxidation of the lignin molecule at sites remote from the enzyme, and eventually cause the degradation of lignin.

Dehydrogenase is an enzyme common to most micro-organisms, but in contrast to the other enzymes discussed in section 2.5 it is predominantly endocellular. The amount of dehydrogenase activity in soil has been used as an indicator of the activity of the soil microbial population. 2,3,5-triphenyltetrazolium chloride (TTC), a pale-coloured water-soluble compound, acts as an electron acceptor for many dehydrogenases and is reduced to triphenyl formazan, a red-coloured water-insoluble compound soluble in methanol, which can be measured colorimetrically.

2.6 Measurement of microbial biomass

Dehydrogenase activity is an example of a biochemical process which has been used to estimate the activity of the soil microbial biomass. Other biochemical properties of soils have been used to estimate both the size and the activity of the microbial population.

2.6.1 *ATP content*

ATP is a compound common to all organisms, and if it assumed that all micro-organisms contain the same proportion of ATP, then the amount of ATP present in soil can be related to the amount of microbial biomass (assuming larger organisms are absent from the soil sample). ATP is extracted from soil using paraquat and tri-chloro-acetic acid, and measured by the amount of light produced when it reacts with luciferin-luciferase (obtained from firefly tails). A correction factor is used to allow for the efficiency of extraction of ATP from soil (section 2.6.2).

The soil microbial population appears to maintain a high ATP content despite the limited substrate supply to these organisms (section 2.1.2). This supply may be supplemented by *cryptic growth* (utilisation of dead cells as substrates), metabolism of endogenous energy reserves (such as poly-hydroxybutyrate) or decomposition of more resistant fractions of the soil organic matter.

2.6.2 *Chloroform fumigation*

If a soil sample is fumigated with chloroform most of the microbial population is killed. The amount of carbon, nitrogen, phosphorus and sulphur released from this biomass following lysis can be measured and related to the amount of living microbial biomass originally present in the soil.

There are two main variations on this basic method, fumigation–incubation and fumigation–extraction. In the *fumigation–incubation* method the fumigated soil, after the chloroform has been removed, is inoculated with fresh soil. The colonising microbial population decomposes the killed microbial biomass and the amount of carbon dioxide produced over a period of 10 days is used to estimate the amount of carbon in the original microbial biomass. An unfumigated control is used to allow for decomposition of non-living soil organic matter by the colonising population. Not all of the killed biomass is decomposed by the colonising population and a correction factor has been obtained by comparing the data obtained from this method with estimates of biomass obtained by direct counts (a laborious and highly skilled method unsuitable for routine biomass measurements).

In the *fumigation–extraction* method the soil is extracted immediately after fumigation and the amount of organic carbon or nitrogen in the extract is determined. The amount of amino-nitrogen in the extract may also be measured using the ninhydrin reaction and used to estimate the amount of biomass nitrogen in the soil. This method and the other methods described in this section have all been calibrated against the fumigation–incubation method to obtain the appropriate correction factors.

2.6.3 Substrate-induced respiration

The final method for measuring microbial biomass is a more physiologically-based method. The rate of respiration (carbon dioxide production) by the microbial population is measured following the addition of substrate to the soil, but before population growth occurs. For each soil used, the minimum quantity of substrate (glucose) required to give the maximum respiration response must be determined. The data obtained reflect the potential activity of the microbial population, but may also be used to estimate microbial biomass when calibrated against the fumigation–incubation method (section 2.6.2).

CHAPTER THREE
ROOT–MICROBE INTERACTIONS

In the previous chapter the loss of material by roots was considered as a source, perhaps the major source, of substrates for soil micro-organisms (section 2.1). It is not surprising therefore that microbial activity is stimulated in the soil around roots, and that these organisms have both beneficial and detrimental effects on plant growth and development. Some micro-organisms form symbiotic relationships with roots, for example root nodules and mycorrhizas. This chapter examines the interactions between micro-organisms and plant roots in more detail.

3.1 The rhizosphere

At the beginning of this century it was observed that bacteria were more abundant in the soil surrounding plant roots than in soil further away from the root; this zone of soil was termed the *rhizosphere*. This definition has more recently been elaborated to distinguish the surface of the root (the *rhizoplane*) and the outer zone of the root itself (*the endorhizosphere*) (Figure 3.1). In addition to exerting direct biological effects, roots may also affect the chemical and physical properties of the soil, thereby indirectly influencing soil micro-organisms (section 4.10).

The stimulation of microbial activity around plant roots is primarily due to the leaky nature of roots which provide substrates for microbial growth (section 2.1.1). Only 5–10% of the root surface appears to be covered by micro-organisms when viewed under the light microscope. This is probably due to the small number of points of direct contact between sources of inoculum (for example, organic debris) and the root surface. The distribution appears irregular, with small clumps and larger aggregates often associated with cell junctions and moribund cells, reflecting the supply of substrates from the root (Figure 1.3c).

Organic materials in the rhizosphere have been classified according to

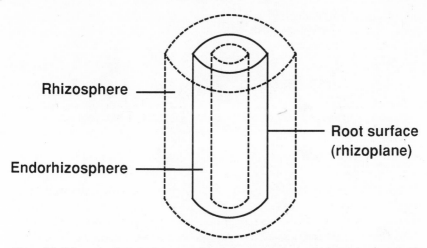

Rhizosphere

Root surface
(rhizoplane)

Endorhizosphere

Figure 3.1 Schematic diagram (not to scale) of the zones of microbial activity associated with a plant root.

their origin in the root (Figure 3.2), although they are not always well-defined compounds from a single origin. *Exudates* are low-molecular-weight compounds which leak from all cells either into intercellular spaces and then into the soil through cell junctions, or directly through epidermal cell walls into the soil. *Secretions* comprise both low-molecular-weight compounds and high-molecular-weight mucilages which are released actively by the plant from the root cap and from epidermal cells. There are *mucilages* produced by bacterial degradation of the outer primary cell wall of dead epidermal cells. *Mucigel* is a layer of gelatinous material on the root surface which is visible under the transmission electron microscope (Figure 1.3c); it comprises natural and modified plant mucilages, bacterial cells and their metabolic products such as capsules and slimes, and also colloidal soil mineral and organic matter. Finally, *lysates* are compounds released from the autolysis of older epidermal cells following failure of the plasmalemma. These cells are then colonised by micro-organisms causing further release of materials into the rhizosphere.

Little is known about the kinetics of microbial growth in the rhizosphere. Rather than being continuously fed with substrate the bacteria in the rhizosphere probably receive substrate in batches. Most studies on growth kinetics are carried out in stirred solution, but growth in soil is more likely to occur at surfaces where growth rates and other parameters can be quite different from those in solution. Generation times have been reported for

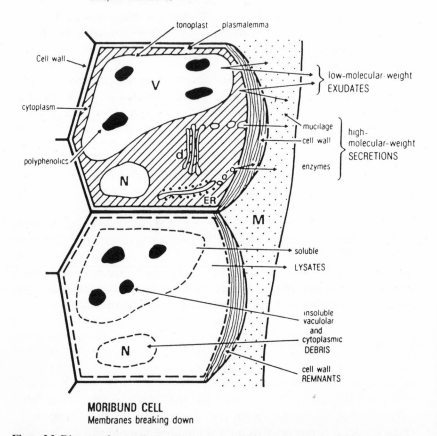

LIVE CELL
Complete membranes

tonoplast   plasmalemma

Cell wall

V

low-molecular-weight
EXUDATES

cytoplasm

mucilage
cell wall

high-
molecular-weight
SECRETIONS

polyphenolics

d

enzymes

N

ER

M

soluble
LYSATES

insoluble
vaculolar
and
cytoplasmic
DEBRIS

N

cell wall
REMNANTS

MORIBUND CELL
Membranes breaking down

Figure 3.2 Diagram of root cells showing the sources of the major root-derived material found in soil (*ER* endoplasmic reticulum, *M* Mucigel, *N* nucleus, *V* vacuole). (After R.C. Foster and J.K. Martin (1981), in *Soil Biochemistry* Vol. 5, eds. E.A. Paul and J.N. Ladd, Marcel Dekker, New York, 75–111.)

Pseudomonas sp. and *Bacillus sp.* on roots of *Pinus radiata* of 5 h and 39 h respectively. These growth rates are slower than for bacteria in laboratory media. Pseudomonads show faster rates of growth in the rhizosphere than do bacilli; this relatively rapid rate of growth by pseudomonads around plant roots makes them useful organisms for controlling root pathogens (section 6.3.2).

Mathematical modelling of microbial growth in the rhizosphere has predicted large changes in microbial numbers over distances of a few millimetres from the root surface. The extent of the rhizosphere depends upon conditions such as the soil moisture content. Such gradients in microbial numbers around roots are due to both the diffusion of low-molecular-weight and volatile compounds away from the root and the utilisation of root-derived material and the release of by-products by micro-organisms. By making several major assumptions it is possible to predict the microbial biomass in the rhizosphere of barley (*Hordeum vulgare*) growing under field conditions using laboratory measurements of root loss. Predicted biomass values of $41–168\,g\,C\,m^{-2}$ agree well with measured biomass values of $43–162\,g\,C\,m^{-2}$. This supports the hypothesis that the rhizosphere is the major zone of microbial activity in soil.

A major problem in testing these predictions is the lack of satisfactory techniques for measuring microbial numbers or biomass over short distances around roots in soil. The traditional method has been to remove a plant from soil, shake off loosely adhering soil, and then estimate, using plate counts, the number of bacteria or fungi per gram of 'firmly adhering' (rhizosphere) soil. The ratio of this number to the count per gram of non-rhizosphere soil gives an R:S ratio. Such data (Table 3.1) suggest a stimulation of growth of specific groups of organisms in the rhizosphere, for example, denitrifiers with an R:S value of 1260 (section 3.2.2).

3.2 Microbial processes in the rhizosphere

The release of a high proportion of carbon fixed during photosynthesis (*photosynthate*) from roots into the soil is a major factor in the carbon

Table 3.1 The numbers of different groups of micro-organisms (on a soil dry weight basis) in the rhizosphere (R) of wheat (*Triticum aestivum*) and in non-rhizosphere (S) soil, and the calculated R:S ratio. After J.W. Rouatt et al. (1960) *Proceedings of the Soil Science Society of America* **24**, 271–273.

Micro-organisms	Rhizosphere (no. g^{-1})	Non-rhizosphere (no. g^{-1})	R:S ratio
Bacteria	120×10^7	5×10^7	24.0
Fungi	12×10^5	1×10^5	12.0
Protozoa	24×10^2	10×10^2	2.4
Ammonifiers	500×10^6	4×10^6	125.0
Denitrifiers	1260×10^5	1×10^5	1260.0

Table 3.2 Microbial processes in the rhizosphere which may influence the plant.

Beneficial
  Associative nitrogen fixation
  Production of plant growth hormones
  Increase in availability of nutrients
  Competition with root pathogens
  Production of mucilage

Detrimental
  Root pathogens
  Immobilisation of nutrients
  Production of phytotoxins
  Denitrification

economy of plants, and some benefit to the plant might be expected. Some microbial processes do appear to be stimulated in the rhizosphere (Table 3.2), although the benefit to the plant is not always obvious.

3.2.1 *Associative nitrogen fixation*

Non-symbiotic nitrogen-fixing bacteria are common in the rhizosphere and under conditions of nitrogen limitation may have a competitive advantage over other micro-organisms. There is no clear evidence in support of selective stimulation of these organisms, and these bacteria do not appear to invade the root. However, specific associations between *Azotobacter paspali* and *Paspalum notatum* (a tropical grass) and between *Azospirillum sp.* and cereal roots have been suggested. The extent of rhizosphere-associated nitrogen fixation depends upon the supply of oxidisable carbon and the efficiency of its conversion. Nitrogen fixation is costly in terms of its ATP requirement and measured efficiencies for free-living organisms range from $4\,\mathrm{g\,C\,g^{-1}\,N}$ in *Azospirillum brasiliense* to $174\,\mathrm{g\,C\,g^{-1}\,N}$ in *Aerobacter aerogenes*. Only simple carbon compounds can be utilised. If it is assumed that the efficiency of conversion of root-derived carbon is $10\,\mathrm{g\,C\,g^{-1}N}$, that the nitrogen-fixing bacteria comprise 10% of the rhizosphere population, and that all of the carbon lost from roots is equally available for use by all bacteria, then if $150\,\mathrm{g\,C\,m^{-2}}$ are lost from the root (as estimated for a wheat crop) the maximum potential nitrogen fixation is only $1.5\,\mathrm{g\,N\,m^{-2}}$.

Evidence of a significant role for rhizosphere-associated nitrogen fixation comes from two sources. Nitrogen balance sheets for different vegetation systems where legumes are absent often show an accumulation of nitrogen in excess of the amounts lost by plant removal, leaching and denitrification

Table 3.3 Net gains of nitrogen for soil under different vegetation systems. After Giller and Day (1985).

Vegetation	Location	N balance $(g\,m^{-2}\,yr^{-1})$
Grassland		
Kentucky bluegrass	USA	1.7
Dactylis glomerata	UK	7.8
Agricultural		
Wheat	UK	4.1
Rice	Thailand	4.0
Woodland		
Mixed woodland	UK	5.3
Highland rainforest	Ghana	1.7–4.5

(Table 3.3). This could be due to inputs from rainfall of the order of $1.5\,g\,N\,m^{-2}\,yr^{-1}$ but possibly higher (section 4.3.3), and from nitrogen fixation either by cyanobacteria (section 1.3.2.4) or by rhizosphere bacteria. Measurements of nitrogen fixation in the rhizosphere of cereals and grasses using the acetylene reduction assay (section 2.4.1) have indicated rates of fixation as high as $0.2\,g\,N\,m^{-2}\,day^{-1}$. However, excision of roots can stimulate loss of root material, which, coupled with long delays between sampling and exposure of both excised roots and intact soil cores to acetylene, can lead to overestimates of rates of nitrogen fixation. The inhibition of ethylene oxidation (a normal soil process) by acetylene causes further large and variable overestimates. Many of the early results on rates of associative nitrogen fixation must, therefore, be viewed with caution.

Incorporation of isotopically enriched $^{15}N_2$ gas into tropical grasses and cereals under laboratory conditions demonstrates the presence of active nitrogen-fixing organisms in the rooting medium and the absorption of some fixed nitrogen by the plants. However, fixed nitrogen will be in the form of organic nitrogen and will only become available to the plant following mineralisation to NH_4^+ or NO_3^-. It seems likely, therefore, that the contribution from associative nitrogen fixation in the rhizosphere is only small. The significance of this input in terms of plant productivity remains unclear.

3.2.2 Changes in nutrient availability

Under conditions of low nutrient supply, the rhizosphere population may compete for certain nutrients, thereby reducing the supply to the plant.

However ^{32}P tracer techniques with barley (*Hordeum vulgare*) growing in solution culture indicate that the amount of phosphate available to the plant can be either reduced or increased compared with a sterile root system. Manganese in the form of Mn^{2+} may be oxidised to insoluble manganese dioxide in the rhizosphere causing manganese deficiency in oats, although in solution culture the rhizosphere bacteria produce compounds (*ionophores*) which promote manganese uptake by roots.

Immobilisation of nutrients may be encouraged in the rhizosphere by the high C:N ratio of root-derived material. This may be useful for mobile nutrients such as NO_3^- which might otherwise be leached from the rooting zone. Furthermore, a high proportion of soil bacteria are facultative anaerobes and respiration by both the root and the rhizosphere population may reduce the redox potential sufficiently to allow denitrification (section 2.4.3) to occur. However, evapotranspiration may cause the rhizosphere soil to become drier (section 4.10) thereby increasing the rate of oxygen diffusion, and uptake by the plant may remove nitrate from the potential denitrifying zones. Oxygen availability may also be increased via aerenchyma cells within the plant allowing diffusion to the rhizosphere, particularly in aquatic plants. There are, therefore, several possible fates of NO_3^- in the rhizosphere (Figure 3.3).

3.2.3   *Production of plant growth hormones*

Rhizosphere micro-organisms produce compounds such as *growth hormones* and *phytotoxins* which may affect plant growth. The diversity of substrates available for growth in the rhizosphere means that many products are likely. It is relatively simple to demonstrate the production of a particular compound by an organism and its effect on a plant in the laboratory, but it is more difficult to demonstrate that the compound is biologically active in the form and at the concentration at which it is present in soil. Measurement in soil is difficult when the compound is present in low concentrations and is produced at localised sites. Identification of a compound is usually by bioassay, which can only lead to descriptions such as 'auxin-like'. Some compounds, for example auxins and ethylene, inhibit plant growth at one concentration but stimulate growth at a lower concentration. Most of the major types of plant hormones can be produced by bacteria and fungi.

Indole acetic acid is an auxin which is produced from tryptophan. It has been implicated in the curling of root hairs on legume roots in the presence of the appropriate rhizobia (section 3.4.2). It is also metabolised by soil

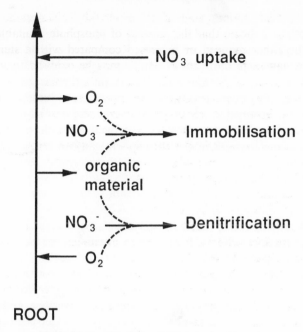

ROOT

Figure 3.3 Possible fates of NO_3^- in the rhizosphere.

bacteria, therefore its significance in soil will depend upon its rate of accumulation. Under anaerobic conditions *ethylene* can occur in the soil atmosphere at concentrations sufficient to inhibit extension of cereal roots in the laboratory. It is produced from methionine and is also metabolised by micro-organisms. Little information is available on cytokinins and abscisic acid in soil.

The phytotoxic effects of one crop on the succeeding one have been observed for centuries and it is now assumed that microbial breakdown of plant residues is involved. When the effect of a micro-organism is on seed germination it is not easy to distinguish a metabolite effect from a physical effect such as reduced oxygen concentration. Early work suggested that dihydroxystearic acid and vanillin were major phytotoxins, but aliphatic acids such as acetic and oxalic acids are also produced by bacteria and fungi, particularly under anaerobic conditions which also prevents their degradation. Substrates are derived from soil organic matter, green manures, crop residues and animal slurries. Less is known of the effects of aromatic acids such as *p*-hydroxybenzoic acid and *p*-coumaric acid which have been detected in soils. Many of these phenolic compounds eventually form part of the humic acid fraction of the soil organic matter (section 1.2.2).

Antibiotic production by soil micro-organisms occurs in laboratory culture, but production of these compounds in soil has yet to be demonstrated. Antibiotics may inhibit or stimulate plant growth either directly, or indirectly by the removal of pathogens. If produced, they can be inactivated by adsorption to clay or degraded by micro-organisms. Sulphur-containing volatiles such as hydrogen sulphide produced by sulphate-reducing bacteria in the rhizosphere can cause toxicity to rice (section 2.4.5).

3.2.4   Allelopathy

There are many reports of inhibitory or stimulatory interactions between plants, and between plants and micro-organisms (including those described above). These interactions have been termed *allelopathy*. For example, couchgrass (*Agropyron repens*), a weed found in many countries, decreases growth of wheat (*Triticum aestivum*) and lucerne (*Medicago sativa*). Vegetation under black walnut (*Juglans nigra*) is very sparse, and this has been attributed to a toxic substance similar to juglone which is produced by walnut trees and is toxic to adjacent plants. In the UK heather (*Calluna vulgaris*) reduces growth of Sitka spruce (*Picea sitchensis*) but not Scots pine (*Pinus sylvestris*).

Concentrations of NO_3^- in cropped soil are often lower than in uncropped soil, even after taking into account the nitrogen taken up by the crop plants and leaching losses (which are lower under cropped soils than fallow soils, section 6.2). There is some evidence that nitrification is affected by allelochemicals produced by many plant species and micro-organisms. There is also evidence that the role of nitrification is increasingly reduced as plant succession moves towards the climax vegetation (section 5.4.3). The mechanisms for many of these allelopathic interactions remain unclear.

3.3   Plant root pathogens

Roots growing through soil are exposed to infection by pathogens, particularly by fungi. Infection is more likely if the root is damaged by abrasion with soil particles or by the emergence of lateral roots. However, as with symbionts such as *Rhizobium* and mycorrhizal fungi, potential pathogens must compete with the rhizosphere microflora for substrates for growth and infection sites. As only a small proportion of the surface area of young roots is covered by bacteria, this affords little protection against pathogens which move to the root and infect rapidly, for example *Phytophthora cinnamomi*.

The ability of a pathogen to infect a root depends upon the size of the fungal propagules (an indication of the reserves for growth through soil), the number of propagules, the rooting intensity of the plant and the ability of the pathogen to overcome the resistance of the host to infection. Organisms with smaller propagules need to produce more of them to come near enough to a root for colonisation. Plant species with low rooting densities may avoid infections by pathogen populations likely to give many infection points on a species with a higher rooting density.

Although fungi prefer to grow along cell junctions, the ability of fungi to translocate nutrients allows them to grow across areas of low substrate supply. This may also allow some escape from antagonistic bacteria. However, roots will normally outgrow fungi and, therefore, lateral colonisation from soil remains an important means of infection under conditions favourable for root growth.

The fungal flora of the root changes as the root ages. Many of the fungi which develop on young roots can be weak unspecialised pathogens, entering and killing juvenile but not mature roots. The attack is usually only serious if seedling growth is slowed or stopped, resulting in damping-off of seedlings growing under unfavourable conditions. The specialised vascular wilt fungi (*Fusarium sp.* and *Verticillium sp.*) can also only attack juvenile roots, but once inside the root they escape into the vascular system.

The rhizosphere may contain fungi such as *Trichoderma viride* which are antagonistic to the root infecting fungi (see Chapter 6 for details of biological control of root pathogens). It is sometimes possible to weaken the root rot fungus during its resting phase in soil. For example, the incidence of banana wilt in land heavily infested with *Fusarium oxysporum* f. *cubense* can be reduced by waterlogging the soil for several months.

The specialised root rot fungi are able to attack mature root cells and decompose lignocellulose. They often grow along the root surface before penetrating the root cells. In undisturbed equatorial forests *Armillaria mellea*, a vigorous pathogen, grows over the surface of roots of some tree species as a weft without penetrating the roots, provided the tree is healthy. However, under conditions of severe drought, or if the tree is cut down, the fungus enters and grows inside the root. Such roots can carry the fungus for several years and are potential sources of infection for any new susceptible crop growing in that soil. This may be important when natural forest is converted to plantations of crops such as tea (*Camellia sinensis*) or rubber (*Hevea brasiliensis*).

3.4 The legume-*Rhizobium* symbiosis

Certain members of the rhizosphere community can enter into a symbiotic relationship with the host plant forming mycorrhizas (section 3.7) and root nodules. Nodules are formed on the roots of actinorhizal plants by *Frankia sp.* (section 3.5), and on the roots of legumes by *Rhizobium sp.* and *Bradyrhizobium sp.* The bacteria obtain carbohydrate from the host plant and supply the plant with nitrogen compounds derived from atmospheric nitrogen. Legumes are important components of natural plant communities and agricultural systems. Most of the available information is on agricultural legumes. The inoculation of legumes with *Rhizobium* is discussed in section 6.3.1.

3.4.1 *Role of legumes in agricultural systems*

Symbiotic nitrogen fixation by leguminous plants provides the largest input of nitrogen into natural ecosystems. In New Zealand, where no nitrogen fertiliser is used for pasture production, rates of fixation by white clover (*Trifolium repens*) of over $60 \, g \, N \, m^{-2} \, yr^{-1}$ have been reported (section 2.4.1). The benefits of legumes have been appreciated for centuries; clovers (*Trifolium sp.*) formed a part of the Norfolk four-course rotation system practised in England during the seventeenth century, and mixed cropping systems of beans (*Phaseolus vulgaris*) and maize (*Zea mays*) are common in South America and East Africa today (section 6.5.1).

The great majority of leguminous genera and species are tropical, and the ancestral niche was probably on leached soils in tropical rain forests. The sub-tropical and temperate legumes developed from these. Some of them, particularly those belonging to the Trifolieae such as *Trifolium sp.* and *Medicago sp.* and the Viceae such as *Pisum sp.* and *Cicer sp.* have become adapted to neutral or calcareous soils of higher nutrient status. A high proportion of legumes cultivated in temperate agriculture, which have received most attention, belong to these groups and the role of legumes in other regions has been less extensively studied.

Legume crops can be broadly divided into those grown for their protein-rich seed (grain legumes) such as soybean (*Glycine max*) and chickpea (*Cicer arietinum*), and forage or pasture legumes, such as white clover (*T. repens*) and kudzu (*Pueraria phaseoloides*). Soybeans are currently the dominant grain legume crop with annual production of 10^8 tonnes worldwide.

Nodules can only fix nitrogen actively if the plant is adequately supplied with all the elements essential for growth. Legumes have an additional high requirement for molybdenum, a constituent of nitrogenase (section 2.4.1). Legumes are often grown in acid soils (section 4.3), particularly in tropical and sub-tropical regions, where factors such as high concentrations of H^+, Al^{3+} and Mn^{2+}, and low concentrations of Ca^{2+} and phosphate may limit growth, nodulation and nitrogen fixation. Legumes relying on nitrogen fixation are generally more sensitive to acidity than the same legumes utilising NH_4^+ and NO_3^- (*mineral N*). An effectively nodulated legume (section 3.4.2) may grow well in a soil with a low mineral nitrogen content; however, high concentrations of mineral nitrogen inhibit nodulation and fixation of atmospheric nitrogen, thereby eliminating the advantage associated with a legume.

3.4.2 Infection

Nitrogen fixation in legumes depends upon a highly coordinated sequence of interactions between plants of the family Leguminoseae and soil bacteria belonging to the genera *Rhizobium* and *Bradyrhizobium* which results in the formation of root nodules. Only 48% of the leguminous genera have been examined for nodules to date, and of these 86% were found to be nodulated. Not all legumes are nodulated by all rhizobia. For example, lucerne (*Medicago sativa*) rhizobia only nodulate lucerne and no other legumes, and rhizobia that nodulate other legumes do not nodulate lucerne. The grouping of legumes nodulated by the same species of bacterium gave rise to the concept of *cross-inoculation specificity*.

In *Bergey's Manual of Systematic Bacteriology* (1984) the original classification of nodule bacteria as a single genus, *Rhizobium*, divided into species according to their cross-inoculation specificity was modified to include two genera. *Rhizobium* is now classified as a genus of fast-growing species producing acid in laboratory media, which nodulate mainly temperate legumes. The new genus, *Bradyrhizobium*, comprises slow-growing species producing alkali in laboratory media, which nodulate mainly tropical legumes.

In the non-symbiotic state rhizobia are common Gram-negative, non-spore-forming rod-shaped soil bacteria, living saprophytically on a wide range of organic carbon sources, but unable to fix nitrogen. In the presence of the appropriate legume root they are able to overcome the plant's defence

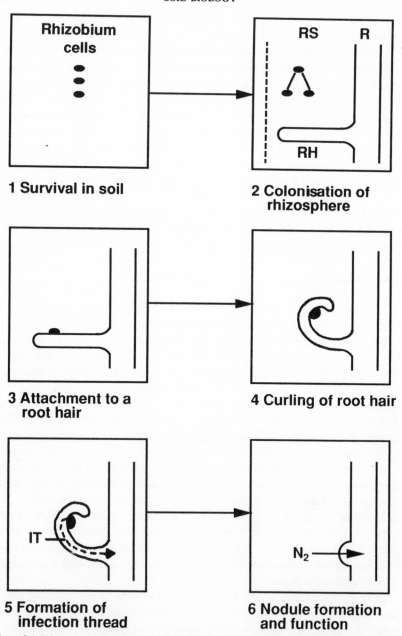

Figure 3.4 Schematic diagram of the stages leading to nodulation in most legumes (*R* root, *RS* rhizosphere, *RH* root hair, *IT* infection thread).

mechanisms, infect the root, form nodules, synthesise nitrogenase and other vital components and fix nitrogen.

Despite a great deal of research into the infection of legumes by rhizobia the detailed mechanisms remain unclear. The major stages are shown in Figure 3.4. There is no convincing evidence that legume root exudates specifically stimulate multiplication or chemotaxis (section 4.7) of the appropriate rhizobia. However, multiplication in the rhizosphere is necessary if the rhizobia are to compete with other rhizosphere organisms for potential infection sites on the root surface. Curling and branching of root hairs is the first visible response by the host legume, possibly due to the production of indole acetic acid by rhizobia from tryptophan excreted by the roots. A marked degree of curling by root hairs (360° or more) appears to be characteristic of a compatible interaction.

Attachment of rhizobia to the root surface may be one of the critical recognition stages prior to successful infection. It has been proposed that a unique immunologically cross-reactive antigen (a glycoprotein lectin) on the cell walls of white clover (*T. repens*) and *R. leguminosarum* biovar *trifolii* is responsible for attachment. A second stage of adherence is suggested which firmly anchors the rhizobia to the root hair surface, possibly by the production of extracellular cellulose microfibrils. Attached rhizobia are often enclosed as the root hair curls, and this may lead to a localised increase in the concentration of chemical signals from the rhizobia and subsequent responses from the plant.

Only a small and variable proportion of root hairs become infected, and many infections (68–99% in *Trifolium sp.*) abort before they reach the base of the root hair. *Infection threads* are tubular structures that carry the rhizobia, often in single file, from the point of entry into the root hair to the inner cells of the root cortex. Rhizobia may cause growth of the host cell wall to be re-orientated so that the root hair becomes invaginated. Alternatively, the rhizobial exopolysaccharide may stimulate an increase in plant pectic enzyme (polygalacturonase) activity which causes a softening of the root hair cell wall and allows penetration by rhizobia. Infection thread formation can be likened to a controlled incompatible host-pathogen reaction.

Not all legumes are infected via root hairs. In peanut (*Arachis hypogaea*), many mimosoid legumes, and the stem nodulating *Sesbania sp.*, infection occurs at emerging lateral roots, or directly through the epidermis. Because many legumes have neither root hairs, nor nodules associated with lateral roots, infection between epidermal cells may be common.

3.4.3 Nodule formation

When the rhizobia penetrate the inner cortical cells of the root they multiply and cause a proportion of the cells to start proliferating thereby forming a *nodule*. Once the bacteria have filled a proliferating cell they become enlarged and change into *bacteroids*. They lose most of their ribosomes and the ability to multiply, but synthesise nitrogenase and are surrounded by membranes formed by the host cell. The bacteroids are bathed in a solution containing *leghaemoglobin* which transports oxygen for respiration at very low partial pressures thereby protecting the oxygen-sensitive nitrogenase. Leghaemoglobin, which is similar to haemoglobin, gives nodules a characteristic pink colour. The diffusion resistance of the nodule also restricts the rate of diffusion of oxygen into the nodule.

Nodules may be spherical, cylindrical, flattened and often bidentate or with coralloid branching, or they may be irregular in shape (Figure 3.5). There may be some 10^{10} bacteroids in a nodule, generally containing a single strain of *Rhizobium*, although multiple-strain occupancy does occur. As the nodule ages, the cells lose their leghaemoglobin and a large vacuole appears, followed by necrosis. It is not clear whether the viable rhizobia released into the soil following nodule death are derived from bacteroids or from cells remaining in infection threads which did not form nodules.

Strains of rhizobia which form nodules on a legume are termed *infective* (strictly speaking, non-infective strains of *Rhizobium* or *Bradyrhizobium* do not exist, because the primary classification criterion is the ability of an organism to form nodules). Those strains which then proceed to fix nitrogen at high rates are termed *effective* and those which are poor nitrogen fixers are termed *ineffective*. However, even legumes nodulated by highly effective strains continue to take up a small proportion of nitrogen from soil as NO_3^- and NH_4^+, providing a supply is available. The same strain may exhibit different levels of effectiveness on different host species or varieties. Roots infected by ineffective strains carry more but smaller and paler nodules than those infected by effective strains.

Only certain parts of the legume root are susceptible to infection at a particular time, and root cells in these infection zones remain infectible for only 12–24 h. Incompatibility between strain and host may occur at any of the stages in infection and nodulation, and therefore, cross-inoculation specificity (section 3.4.2) is not necessarily restricted to the early stages of attachment to the root. *Rhizobium* polysaccharides appear to be important in determining legume compatibility; mutants lacking normal exopoly-saccharide show poor infection and nodulation.

Figure 3.5 Nodule formed on the root of lucerne (*Medicago satira*) by *Rhizobium meliloti* (magnification × 3).

In *Rhizobium*, the genes for root hair curling, infection thread formation and host-specific nodule induction are located on large indigenous Symplasmids (section 6.4), which also carry the genes for nitrogenase formation (*nif*) and nitrogen fixation (*fix*). In *Bradyrhizobium*, the functional analogues of these *nod*, *nif* and *fix* genes are carried on the chromosome. Four of the nodulation genes, *nod* D, A, B, C are contiguous and highly conserved. These genes control root hair curling, and *nod* D is a regulatory gene which is *constitutive* (always produced rather than being *inducible*) in *Rhizobium leguminosarum* biovar *trifolii*.

The nod genes are regulated, via nod D, by compounds in the legume root exudates. In white clover, *hydroxyflavones* produced in the infectible zone of emerging root hairs stimulate expression of the nod genes within minutes, whereas *coumarins* and *isoflavones* produced behind the root tip repress *nod* gene transcription. The success of a particular infection depends upon the ratio of these stimulatory and inhibitory compounds, which provides the plant with a mechanism for regulating nodulation. Once *nod* D has interacted with the flavonoids, other *nod* genes such as *nod* F and *nod* E which are involved in host range specificity are induced, and these in turn may regulate other non-Sym plasmid genes, such as those involved in polygalacturonase production.

The amount of nitrogen fixed by a leguminous crop depends upon the effectiveness and longevity of the nodules. Nodules of annual plants tend to die at flowering and seed set, probably due to competition for carbohydrates. The cutting or hard grazing of clovers may also cause the death of nodules as a result of the interruption in photosynthesis. The onset of drought often causes the crop to shed its nodules. Nodules of leguminous shrubs or trees may persist for several years. Nodules may be short-lived due to being parasitised by the larvae of insects such as the pea weevil (*Sitonia lineata*). The severity of this attack can be reduced by irrigation in dry seasons, thereby prolonging the life of the nodules.

3.4.4 Carbon economy of legumes

The bacteria in the nodules must be supplied with energy in order to fix nitrogen, and therefore they also need oxygen for the oxidation of carbohydrate. The nodule and associated root system receive 15–30% of

Table 3.4 Comparison of theoretical carbon costs for the process of nitrogen fixation by legume nodules with nitrogen assimilation from soil nitrate. After Atkins (1984).

Component	Cost $(g C g^{-1} N)$
Biological fixation	
Nitrogenase/hydrogenase	1.7–3.5
Ammonia assimilation and ancillary C metabolism	0.4–0.5
Transport of N	0.3
Growth and maintenance of nodule	0.5–1.8
Total	2.9–6.1
Nitrate assimilation	
Nitrate uptake	0.1
Nitrate and nitrite reductases	0–1.5*
Ammonia assimilation and ancillary C metabolism	0.4–0.5
Transport of N	0.3
Total	0.8–2.4

0–1.5*: 0 = complete reduction in shoots with direct utilisation of photosynthetic reductant; 1.5 = complete reduction in roots at the expense of respired assimilate.

the net photosynthate produced by the legume. Although it is impossible to measure the individual costs of nodule function, theoretical estimates suggest that nitrogenase activity is the major respiratory cost in the nodule (Table 3.4). Part of this cost is due to the reduction of protons to hydrogen gas by the enzyme whilst reducing nitrogen. Some strains of rhizobia possess a hydrogen uptake (Hup) system which recycles the hydrogen by oxidising it to water. Nodules possessing the Hup system should utilise photosynthate more efficiently; however, no clear yield benefit has been demonstrated.

Although nitrogen fixation represents an economy in terms of the costs of agricultural production, a legume fixing nitrogen does pay a price in terms of the respiratory costs involved. A legume obtaining most of its nitrogen from fixation may require more energy for nitrogen assimilation than the same legume utilising mainly soil mineral nitrogen (Table 3.4). However, under field conditions no difference is generally observed between the yields of legumes dependent mainly on fixation, and the same legume utilising soil mineral nitrogen.

The NH_4^+ produced by fixation is excreted into the host cell cytoplasm where it is assimilated and used to synthesize organic nitrogen for transport within the plant. Legumes can be grouped according to the form in which nitrogen is exported from the nodule. Temperate legumes such as clovers (*Trifolium sp.*) and lucerne (*Medicago sativa*) export *amides* such as asparagine and glutamine, whereas tropical legumes such as *Phaseolus sp.* and soybean (*Glycine max*) export *ureides* such as allantoin and allantoic acid. Ureide transport may be more efficient than amide transport in terms of moles of C used in their synthesis.

Despite the importance of legumes in agriculture and natural ecosystems there are no reliable estimates of the amounts of nitrogen fixed under field conditions. Rates of fixation have been reported for many legumes, but these estimates vary considerably, for example, $4.5–67.3 \, \mathrm{g \, m^{-2} \, yr^{-1}}$ for white clover (*Trifolium repens*), and depend greatly upon the method of measurement used (section 2.4.1).

3.5 Actinorhizas

Several genera of perennial woody trees and shrubs form a nitrogen fixing symbiosis with the actinomycete *Frankia*. Actinorhizal nodules occur in 21 diverse genera in eight families, and although these plants rival legumes in their contribution to global nitrogen fixation, our understanding of the biology and physiology of this symbiosis is poor. Actinorhizal plants are pioneer species on sites poor in nitrogen and often exposed to extreme environmental conditions. They are often found on raw mineral soil and along streams, and in high latitude countries such as Canada and Scandinavia actinorhizal plants thrive under conditions which are unsuitable for legumes. *Alnus sp.* is used in reclamation of material such as china clay waste and colliery spoil where trees such as *Picea sitchensis* often benefit from being interplanted with *Alnus sp.*.

Frankia, which was first isolated in pure culture as recently as 1978, is a genus of Gram-positive, filamentous spore-forming bacteria, which form swellings (vesicles) on the tips of the hyphae. Unlike *Rhizobium*, these organisms can fix nitrogen in the absence of the host plant, and it appears that the vesicles, which form in culture and in the nodule, are the site of nitrogenase activity. Infective isolates are available for only about half of the actinorhizal genera and no species names have yet been given. *Frankia* exhibits a degree of cross-inoculation specificity, with three groups of plant genera recognised: the first comprising *Alnus, Comptonia* and *Myrica*; the second comprising *Elaeagnus, Shepherdia* and *Hippophae*; the third comprising *Casuarina*. The factors which determine this specificity are not known.

Actinorhizal nodules, unlike legume nodules, are modified lateral roots. The hyphae penetrate deformed root hairs within a crypt formed at the junction of several root hairs. Modification of the growth of the root hair tip leads to the formation of an infection thread. Intercellular infection which does not directly involve root hairs also occurs. The endophyte proliferates within the host cortical cells which become hypertrophied with large nuclei and many mitochondria. Phenolic compounds which are inhibitory to *Frankia* accumulate in nearby uninfected cells; this may represent a partial defence response by the host plant. Ineffective nodules (section 3.4.3) may form in which there are no vesicles.

There is little information on carbon and nitrogen metabolism within the nodules, although the nitrogenase is similar to that in other nitrogen-fixing organisms. There is little evolution of hydrogen from actinorhizal nodules, indicating the widespread occurrence of an uptake hydrogenase system. The vesicle wall appears to provide a barrier which limits oxygen diffusion into the vesicle and consequently protects the enzyme from oxygen damage. No further protection is needed, although haemoglobins are found in some nodules.

Nodules, which may be several centimetres in diameter, may last for three to four years in the field and new lobes are added each year. Seasonal patterns of nitrogenase activity occur with a peak in the summer when nodules have completed their growth, and a cessation of activity in winter when the leaves of deciduous plants are lost. Spores formed in senescent nodules may form the major inoculum for soil. In the absence of the host plant the nodulation capacity of a soil decreases with time.

3.6   Agrobacterium

Agrobacterium tumefaciens is a species of soil bacterium closely related to *Rhizobium* and commonly found in the rhizosphere. It contains a large tumour-inducing (Ti) plasmid (section 6.4.2) which allows the bacterium to produce neoplastic overgrowths ('crown galls') on susceptible plants. Wounding of the plant is required before a sequence of processes occurs which leads to gall formation. Galls normally occur where secondary lateral roots break out from the underground stem.

The bacteria become attached to the cells of a wounded plant and anchored by cellulose fibrils produced by the bacteria. The virulence (*vir*) genes in the Ti-plasmid are induced by plant root exudates such as acetosyringone which leads to the formation of a linear single strand of T-DNA which contains the genes for synthesis of auxin and cytokinin. The T-DNA is transferred to the host plant as a protein–DNA complex; a replacement strand is produced in the bacterial donor and a complementary strand in the recipient. It is not known how the T-DNA is transported to the plant cell nucleus nor how it becomes integrated into the host genome.

Following transfer of the T-DNA, plant hormones are synthesised and rapid host cell division leads to the formation of a gall. Compounds termed *opines* are synthesised by the plant under the direction of the foreign DNA which are used preferentially as a substrate for growth by *Agrobacterium*. Opines also induce the transfer of the Ti-plasmid to other bacterial cells by conjugation (section 6.4).

T-DNA transfer in crown gall formation is similar to bacterial plasmid conjugation; the mobilisation functions of a naturally-occurring bacterial plasmid required for plasmid conjugation also allow transfer of this plasmid by *A. tumefaciens* to plants. This raises the possibility that plants may have access to the gene pool of certain soil bacteria. It is not known how widespread this ability is amongst bacteria and plants, however, this ability in *A. tumefaciens* makes it a useful vector for the introduction of foreign genes into plants. A non-pathogenic strain of *A. rhizogenes* has been used successfully to control crown gall formation on a wide range of crops (section 6.3.5).

3.7 Mycorrhizas

The roots of most plant species in natural environments and in cultivation form symbiotic associations, termed *mycorrhizas*, with specialised fungi. This fungus–root symbiosis is, therefore, more widespread than the root–nodule symbiosis which is restricted almost exclusively to certain leguminous and actinorhizal species. In nature, 80% of plants have a root system which is really a mycorrhizal system. The fungus obtains carbohydrate from the plant and supplies nutrients, particularly phosphate, to the plant from the soil.

Two main types of mycorrhizas are commonly found: ectomycorrhizas and endomycorrhizas (Figure 3.6). Trees of boreal and temperate forests have *ectomycorrhizas* with a fungal sheath around lateral roots, intercellular penetration of the root cortex, and a mainly external vegetative mycelium. Most herbaceous and graminaceous species of temperate and semi-arid grassland, as well as many tree species in tropical and subtropical regions, have *vesicular–arbuscular mycorrhizas* (VA mycorrhizas). This is the most common type of *endomycorrhiza* (the others are *ericoid, arbutoid* and *orchidaceous*), with internal storage structures (vesicles), intracellular hyphal structures (arbuscules) and external branched single hyphae spreading through soil. In contrast to *Rhizobium sp.*, both types of fungus show little host specificity.

3.7.1 *Ectomycorrhizas*

Ectomycorrhizas are formed mainly by Basidiomycetes, for example *Lactarius sp.* and *Boletus sp.* Infection may arise from existing mycorrhizal roots which act as point inoculum sources. Mycelia fan out into the soil and when they contact an uninfected root hyphae aggregate to form strands. Mycorrhizal fungi colonising young roots are in competition with the rhizosphere microflora. Once established, a sheath forms which causes modification of root exudation. Some of the hyphae grow between the cortical cells to form the *Hartig net*. This provides a large surface area for the interchange of nutrients between the plant and fungus.

The fungus obtains most of its carbon from the tree in the form of sucrose which is hydrolysed extracellularly by the fungus to glucose and fructose

ECTOMYCORRHIZA

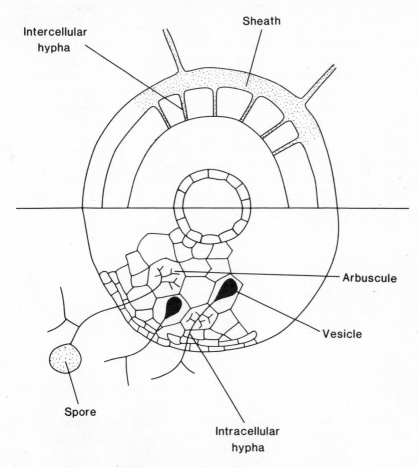

Figure 3.6 A schematic diagram of an ectomycorrhiza and a vesicular–arbuscular mycorrhiza.

(section 2.5.1). These are converted to fungal products (mannitol, trehalose and glycogen) which the tree cannot use, hence a one-way flow is maintained from the host plant to the fungus.

Part of this transfer of carbohydrate may represent carbon which would otherwise be lost from the root, for example, as exudates, but part will

Table 3.5 Carbon fluxes in the rhizosphere of a 35–50 year old Douglas fir stand. After R. Fogel and G. Hunt (1979) *Canadian Journal of Forest Research* **9**, 245–256.

Component	Flux $(g\,m^{-2}\,yr^{-1})$	Percentage of total
Total plant material	3032	100
Mycorrhizal sheath	610	20
Sclerotia and sporophores	232	8
Hyphae	699	23
(Total fungus	1542	51)

constitute an additional demand on the plant's photosynthate supply. A budget for Douglas fir (*Pseudotsuga douglasii*) (Table 3.5) shows that the fungal component comprises half of the annual carbon throughput of the stand. Data for the same stand indicate that 43% of the annual throughput of nitrogen is via the mycorrhizal fungus.

The fungal sheath (Figure 3.6) is of benefit to plants growing in soils low in nutrients because it provides a greater absorbing surface for nutrients than does the root alone. The accumulation of phosphate and other nutrients in the sheath may provide a steady supply of phosphorus, etc. to the plant when transport to the root is reduced, for example during a period of drought. In addition, the relatively rapid turnover of mycelial fungus leads to a concentration of nutrients in the root region. Mycelia are ingested by a variety of small animals such as nematodes (section 1.3.2.6) and Collembola (section 1.3.2.8), and their fruiting bodies and sclerotia are important food for larger animals such as small mammals. Animals, therefore, accelerate the turnover of mycelial tissue and can disseminate mycorrhizal fungal spores.

Increased phosphatase activity of mycorrhizal roots may result in increased hydrolysis of soil organic phosphorus compounds. The fungus may also provide a degree of drought tolerance to the tree. Hyphal strands appear to link plants directly to form a continuous transport network between plants of the same and different species. This may allow the transfer of water, other nutrients, and also carbon, between plants.

3.7.2 Vesicular–arbuscular mycorrhizas

Vesicular–arbuscular (VA) mycorrhizas are formed by fungi belonging to the family Endogonaceae, for example *Glomus sp.* and *Gigaspora sp.*. Roots become infected by hyphae growing through soil from propagules such as spores, or dormant fungal structures in dead previously infected root material, or from nearby roots. The large resting spores (80–150 μm diameter) are produced on the coarse external hyphae, and are borne singly in the soil or aggregated into sporocarps. Spores of *Acaulospora laevis* germinate after 1–2 months, whereas inocula consisting of infected roots can infect rapidly. Spores can germinate in the absence of plant roots; it is not clear whether the presence of roots enhances germination.

The infection process includes arrival of the fungus at the root, penetration and development of the infection, and its spread to other parts of the root. Formation of an *appressorium* (a swollen structure formed on the end of a spore germ tube in contact with the root) often occurs as a prelude to infection. Hyphae then penetrate the epidermal cells or pass between these cells and penetrate the outer cortical cells.

Repeated dichotomous branching of the invading hyphae within the host cell gives rise to an *arbuscule* (Figure 3.6). Between the plasmalemma of the host cell and the wall of the hypha is a polysaccharide matrix. *Vesicles*, 50–70 μm in diameter, form as inter- and intracellular swellings along or at the tips of hyphae. They contain lipid droplets and sometimes glycogen, and probably act as temporary storage organs. The internal structure of VA mycorrhizas can be observed in cleared and stained root samples under the light microscope. Despite the density of internal hyphal development the morphology of the root is little affected by infection. The host cell increases its cytoplasmic content and metabolic activity. In soil containing VA fungus propagules, continued development of the infection is due to spread of existing infections and to new infections.

The reasons for the lack of specificity of VA mycorrhizas are not known. Recognition might occur at the appressoria stage, but these structures are not always associated with infection. The apparent lack of infection of plants belonging to the families Chenopodiaceae, Brassicaceae and Caryophyllaceae could be due to a physical barrier at the cell wall, absence of essential nutrients, or production of toxins by the plant.

Factors such as pH, temperature, pesticides, and nutrient level affect the amount of root infected. For example, increased soil phosphate decreases VA mycorrhizal infection. Phosphate acts on the plant phase and on the soil phase (such as spore germination) of the fungi, and fungi vary in their sensitivity to increased phosphate. Phosphate deficiency causes increased exudation from roots and this may stimulate mycorrhizal infection. To date it has not been possible to grow VA mycorrhizal fungi in pure culture. Growth of the fungus in the symbiotic association may be due to the host plant providing some unknown growth factors in addition to energy sources.

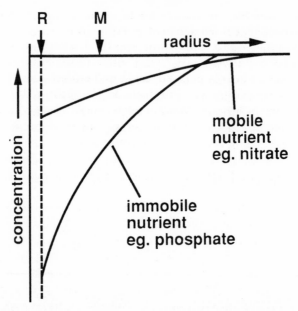

Figure 3.7 The concentration profile for a mobile and an immobile nutrient around a non-mycorrhizal root surface (*R* root surface, *M* indicates position of hyphae if the root were mycorrhizal). (Modified from P.H. Nye and P.B. Tinker (1977) *Solute Movement in the Soil–Root System*, Blackwell Scientific Publications, Oxford.)

VA mycorrhizal plants generally grow better in phosphate deficient soils than non-mycorrhizal plants. Studies using ^{32}P tracers have shown that mycorrhizal plants use the same pool of soil phosphate as non-mycorrhizal plants. The fungus therefore extends the volume of soil from which the plant can take up phosphate rather than solubilising otherwise unavailable forms of phosphorus. This is important for an immobile nutrient such as phosphate which is depleted in the soil around roots, but less important for a mobile nutrient such as NO_3^- (Figure 3.7). Inorganic phosphorus is concentrated by a factor of a thousand in the mycorrhizal hyphae and is stored and transported in the form of polyphosphate. It is not clear how phosphate is transferred into the host plant from the fungus, nor how carbohydrate is transferred to the fungus from the plant.

Mycorrhizal modification of the nutrient uptake properties of roots depends upon the development of extramatrical hyphae in the soil. This can vary from 0.7 m hyphae cm^{-1} infected root for *Glomus mosseae* on onion to 14.2 m hyphae cm^{-1} for *Glomus tenuis* on clover. However, the activity, longevity and turnover time, and hence the significance of these differences in extramatrical hyphal development, is not known.

Mycorrhizas may also increase the uptake of other nutrients such as sulphur, copper, zinc and NH_4^+ and allow plants to grow in harsh conditions such as heavily polluted, acidic and eroded sites. Mycorrhizas sometimes increase resistance to disease and drought and can increase nitrogen fixation by legumes. Many of the beneficial effects of mycorrhizas, for example improved growth in dry soils, may be indirectly due to an improvement in the mineral nutrition of the plant.

3.8 Root-knot nematodes

Root-knot nematodes (*Meloidogyne sp.*) are widely distributed in soils, have an extensive host range and are estimated to account for an annual loss in crop yield of 5% worldwide. Losses due to this pest are particularly large on small farms in developing countries. There are four main pest species: *M. incognita, M. javanica, M. arenaria*, and *M. hapla*. They survive in soil as eggs or larvae, and are disseminated by water, animals and plants.

Larvae develop inside the eggs which hatch, producing second stage juveniles which migrate through soil and are attracted to roots. On contact, the juveniles penetrate the root, using stylets and enzymes, just below the root tip, and move through the root to near the vascular tissue. This causes an increase in plant cell size and galls and knots (1–10 mm in diameter) form on the roots. The juvenile forms change into mature adults, completing their life cycle in 17–57 days. Intensive knot formation causes root deformation, decreased efficiency of uptake of water and nutrients, and premature defoliation and death of plants.

CHAPTER FOUR
SOIL AS AN ENVIRONMENT FOR ORGANISMS

In Chapter 1 the major groups of organisms found in soil, and soil components were described. Soil is probably the most complex and least well understood of ecosystems. Two of the most striking features of the soil population are firstly the enormous diversity of species, and secondly the low rates of growth of soil organisms in relation to their potential growth rates (section 2.1.2). This chapter examines the main physical and chemical factors which determine the occurrence and activity of organisms in soil.

4.1 Temperature

Organisms are often exposed to a wide range of fluctuating temperatures in soil. Soil temperature depends upon atmospheric temperature and inputs and losses of radiation. It is also influenced by the presence or absence of vegetation, the soil water content and the depth within the soil. Temperatures vary between day and night (*diurnal variation*) and throughout the year. The temperature of the soil surface may reach 60 °C or may be below 0 °C. Diurnal fluctuations of up to 35 °C can occur at the soil surface. With increasing depth, diurnal temperature fluctuations are reduced. The mean fluctuation in summer in surface soils in cool temperate climates may be only 10 °C and in sub-tropical and tropical regions may be 30 °C.

Micro-organisms can be categorised according to the range of temperatures at which they grow. *Psychrophiles* grow at 0 °C and have an optimum temperature at or below 20 °C. *Thermophiles* have a maximum temperature for growth of over 50 °C, and a minimum of over 20 °C. *Mesophiles* have temperature optima between these two extremes. The ecological significance of these different responses to temperature is not clear. For example, thermophilic bacteria and fungi are found in temperate soils which appear more suited to psychrophilic organisms.

The ability of micro-organisms to grow at low temperatures may depend upon membrane permeability and associated control of solute transport into the cell. Psychrophilic bacteria appear to synthesise increased quantities of enzymes at low temperatures to compensate for the reduction in the rates of metabolic process at low temperature. It is often assumed that soil microbial activity is negligible at temperatures less than 5 °C, although there are reports of significant denitrification occurring in soils at such low temperatures.

High temperatures in soils are often associated with dry conditions (section 4.6), and the resulting effects on soil organisms and processes can be complex. Figure 4.1 shows the effect of climate on soil respiration in three contrasting ecological zones. In temperate grassland soils, microbial activity is limited by low temperatures during the winter, whereas the major limitation to soil organisms in tropical savannah soils during the same period is drought. In tropical rain forest neither temperature nor soil moisture limit microbial activity throughout the year, leading to relatively constant and rapid rates of organic matter turnover in these soils (section 5.3).

Fluctuations in soil temperature and moisture may have a more profound effect on soil microbial processes than constant extreme conditions. For example, air-drying soil kills some of the microbial population and renders soil organic matter more decomposable. The release of soluble organic carbon, nitrogen and phosphorus following air-drying of soil leads to a subsequent increase in respiration and nitrification when the dry soil is re-wetted (this is known as the *Birch effect*). In tropical areas with distinct wet and dry seasons some crops are planted toward the end of the dry

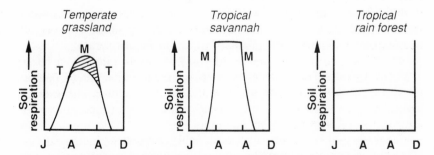

Figure 4.1 Seasonal soil respiration rates for three ecological zones, showing the main climatic limitations (*T* temperature limited, *M* moisture limited). (After M.J. Swift *et al.* (1979) *Decomposition in Terrestrial Ecosystems*, Blackwell Scientific Publications, Oxford.)

season to utilise the flush of mineral nitrogen that occurs at the onset of the rains. The effect of freezing and thawing on soil organisms and mineralisation processes are generally less pronounced than drying and re-wetting.

4.2 Salinity

Soluble salts (mainly sodium sulphate and chlorides of sodium and calcium) accumulate in the surface of soils under hot dry conditions when the groundwater comes within a few metres of the soil surface (as happens after forest clearance) to form *saline soils*. During dry periods, when evapotranspiration is greater than rainfall, the surface of saline soils is covered with a salt crust which dissolves in the soil solution each time it is wetted. Saline soils may also be produced by poor irrigation practice. Irrigation water always contains some dissolved salts, and these salts remain in the soil after the water has evaporated or been transpired by the plant. It is important therefore to ensure that salts are leached from the root zone using water in excess of that required by the crop, and to ensure that drainage is provided.

If the salts are mainly sodium, then as they are washed out, sodium hydrogen carbonate is formed which causes the soil pH value to increase to > 9. In these *sodic soils* the sodium and high pH cause humic colloids to deflocculate and clays to swell or disperse, as the normal attractive forces become forces of repulsion, leading to an unstable soil structure.

Salinity causes dwarfed and stunted plants; this may be due to a decrease in water availability because of the decreased water potential (increase in osmotic pressure) caused by the high salt concentration (section 4.6). The effect of salts in decreasing the water potential may cause symptoms of drought. Some plants adapted to saline soils are better able to extract water from dry soils, but this is not always the case. For example, the coconut palm (*Cocos nucifera*) is salt tolerant but not drought tolerant.

In addition to this *osmotic effect*, salts may also exert a *specific ion effect*. Many fruit trees are sensitive to high concentrations of Na^+ and Cl^- ions, which may interfere with metabolism and nutrient uptake. However, in general the effect of salts on plants depends upon the total concentration of soluble salts rather than on specific ions.

Plants tolerant of salinity must be capable of either excluding salts or adjusting to osmotic stress. The latter is achieved by the uptake and accumulation of inorganic ions such as K^+, or the synthesis of low-molecular-weight organic acids such as malate. Plants are generally more sensitive during the seedling stage. Date, cotton (*Gossypium sp.*) and barley (*Hordeum vulgare*) are examples of salt tolerant plants, whereas beans (*Phaseolus vulgaris*) and white clover (*Trifolium repens*) are sensitive.

Soil micro-organisms face similar problems to plants in saline soils. Although many biochemical functions require particular inorganic ions, an increase in concentration above normal intracellular concentrations may lead to disruption of cell function by reducing enzyme activities. In order to prevent osmotic dehydration and to maintain optimum turgor pressure, most bacteria under osmotic stress accumulate K^+ ions and low-molecular-weight organic compounds such as amino acids, betaine, trehalose and glycerol. Gram-negative bacteria accumulate glutamate using glutamate dehydrogenase (which is sensitive to increase in pH value), and electrical neutrality is maintained by also accumulating K^+ ions. Gram-positive bacteria tend to accumulate proline which does not require K^+ accumulation. The cytoplasmic ionic strength appears to act as a signal which controls the induction of systems for synthesis and accumulation of *compatible solutes*.

The osmotic effect of most organic compounds such as glycerol is lower than that of sodium chloride, and complete osmoregulation may not occur in salt-tolerant bacteria. Other adaptations to saline environments may include changes in cell wall phospholipid composition, and the accumulation of betaines which may stabilise the molecular conformation of various enzymes. Clay minerals may also partially protect micro-organisms against osmotic stress (section 4.9).

4.3 Acidity

Acid soils occur widely around the world, particularly in humid tropical regions, for example the cerrados of Brazil, where soils have been strongly leached over a long period of time. The natural processes of weathering cause soils to become more acidic, and the intensity of the acidity, as measured by soil pH, is affected by the type of minerals in the parent rock, climate and vegetation (section 5.4.3). Acidity is generated by carbon dioxide dissolving in the soil solution, and by nitrification (section 2.4.2) and oxidation of sulphur (section 2.4.4). Where fertilisers such as $(NH_4)_2SO_4$ are used, acidity is generated by nitrification of NH_4^+, and bases such as Ca^{2+} and Mg^{2+} are lost from the soil during leaching of the NO_3^- produced. Plant roots (section 4.10.2), plant litter and carboxyl and phenolic groups on humified organic matter also contribute to soil acidity.

Table 4.1 Estimated inputs of acidity into UK soils. After D.L. Rowell and A. Wild (1985) *Soil Use and Management* **1**, 32–33.

Source	$g\,H^+\,m^{-2}\,yr^{-1}$
Natural	
CO_2 in soil pH > 6.5	0.72–1.28
Organic acids in soils and from vegetation	0.01–0.07
Acid rain	
Wet deposition	0.03–>0.10
Dry deposition	0.03–>0.24
NH_3 and NH_4^+ oxidation	0.07
Land use	
Cation excess in vegetation	0.05–0.20
NH_4^+ oxidation and NO_3^- leaching (fertiliser)	0.40–0.60
Oxidation of N and S from organic matter and leaching	0–1.00

The increase in use of fossil fuels in developed countries has led to an additional input of acidity into soils from atmospheric pollutants, mainly sulphur dioxide and various oxides of nitrogen and NH_3, often referred to as acid rain (section 4.3.3). Estimated inputs of acidity into UK soils are shown in Table 4.1.

As soils become more acidic, basic cations are displaced from exchange sites and leached down the soil profile, and exchangeable H^+ ions take their place on the clay minerals and organic matter. Clays with appreciable amounts of exchangeable H^+ are unstable however, and slowly dissolve, releasing aluminium, magnesium and silica. Aluminium becomes the predominant exchangeable cation. Eventually, after prolonged weathering the clay minerals are destroyed and mainly gibbsite, silicates and various iron oxide minerals remain. Soil acidity is therefore different to the acidity of a solution because it is controlled mainly by ion exchange reactions involving both inorganic and organic soil components.

A range of chemical factors are associated with soil acidity. Acid soils contain large amounts of exchangeable aluminium and small amounts of exchangeable bases, particularly Ca^{2+} ions. The solubility of manganese also increases under acid conditions. The equilibrium with the soluble divalent form of manganese is controlled by the redox potential. In very acid soils which are waterlogged high amounts of exchangeable Mn^{2+} and Fe^{2+} may occur. The amount of readily available phosphate may be low in acid soils as phosphate is adsorbed onto the surface of positively-charged oxides and hydroxides of iron and aluminium. These soil acidity factors may have a greater effect on biological activity in soil than does the concentration of H^+ ions. The adverse effects of soil acidity can be overcome by the application of lime to soil.

4.3.1 Aluminium toxicity

Acid soils often contain appreciable amounts of aluminium, due to the ubiquitous nature of this element in soil (it is the third most common element in the earth's crust), and to the increase in solubility which occurs as acidity increases. Aluminium has been shown to be the most important factor related to acidity that affects growth of many plants including lucerne (*Medicago sativa*) in Australia, cotton (*Gossypium sp.*) in the south-eastern United States and coffee (*Coffea sp.*) in Brazil. Surface soils often have lower amounts of exchangeable aluminium than subsoils at pH values below 5, due to the greater affinity of aluminium for adsorption sites on organic matter than on clay minerals, and additions of organic matter to acid soils can often allow plants to grow satisfactorily.

The symptoms of aluminium toxicity are similar in most plants. The roots are stunted and thickened with the lateral roots appearing as peg-like projections. Aluminium may interfere with phosphorus uptake and metabolism within the plant root, it may interfere with mitosis, and it may displace calcium from the root cell-walls preventing cell growth. Some plant species such as tea (*Camellia sinensis*) are able to tolerate high concentrations within the plant by using chelating agents which allow safe transfer of aluminium to the older leaves where it may accumulate in concentrations up to $20\,000\,\mu g\,g^{-1}$.

Little is known of the mechanisms of aluminium toxicity to soil micro-organisms despite the fact that this may be the major factor limiting microbial growth and activity in acid soils. The microbial component of the legume–*Rhizobium* symbiosis is more sensitive to acidity and aluminium than the host plant, and aluminium appears to enter the bacterial cell and interfere with DNA replication. Species of *Rhizobium* vary in their sensitivity to aluminium, and the ability to tolerate aluminium is one of the important properties required of an inoculant for use in tropical regions.

4.3.2 Protected microsites

Bacteria isolated from acid soils when tested under laboratory conditions often do not appear to be tolerant of pH values corresponding to the pH of the soils from which they were isolated. This has been observed for

Streptomyces sp., Arthrobacter sp., Rhizobium leguminosarum biovar *trifolii* and *Nitrobacter sp.*.

An *acidophilic* population (one that prefers acid conditions) may exist in an acid soil, but the isolation procedures used may favour *neutrophilic* organisms (those that prefer neutral conditions). Alternatively, there may be protected microsites in the soil which have a pH value higher than that of the bulk soil. The acidity of the rhizosphere soil can be quite different from that of the non-rhizosphere soil (section 4.10.2) and localised increases in acidity may occur at the surface of clay minerals (section 4.9). Several distinct populations may coexist in the same soil. For example, only neutrophilic species of *Nitrosomonas* were isolated from an acid forest soil in the USA, but both acidophilic and neutrophilic species of *Nitrobacter* were isolated from the same soil.

Figure 4.2 illustrates these two possible situations. A hypothetical crumb of soil is assumed to consist of nine discrete microsites. Both crumbs have the same average pH, but in one there is a protected microsite in which acid-sensitive bacteria can survive, and in the other there is no protection and all the organisms are acid-tolerant. Soil is a highly heterogeneous medium and pH measurements only indicate average values; it is not possible at present to measure localised pH values in soil around plant roots or particles of decomposing organic matter to test the hypothesis of protected microsites.

Bacteria may be able to modify their local environment to produce favourable microsites. For example, it has been proposed that root-nodule bacteria present in acid tropical soils are adapted to those soil conditions because they are often slow-growing organisms (*Bradyrhizobium sp.*) which produce alkali in laboratory media. However, it has not yet been demonstrated that these organisms produce alkali in soil, nor that the

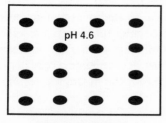

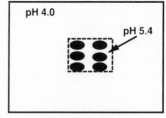

Uniform distribution
Acid-tolerant population
No protected microsites

Clustered distribution
Acid-sensitive population
in protected microsite

Figure 4.2 Hypothetical scheme for the survival of acid-tolerant and acid-sensitive bacteria in an acid soil.

amount of alkali produced would be large enough to bring about a change in pH of the soil surrounding a bacterial cell or colony of cells. Fungi are generally more tolerant of soil acidity than bacteria (Table 5.3), but the reasons for this are not clear.

4.3.3 Acid rain

There is current concern over the effect of atmospheric pollutants, in particular ozone and oxides of sulphur and nitrogen, on plants and soils. This was stimulated by observations of widespread damage to forests in central Europe and in the north-eastern USA. A wide range of species are affected over a large area, and the typical symptoms in Norway spruce (*Picea albies*) are needle discoloration and crown thinning. Increasing damage has been observed in broad-leaved species such as beech (*Fagus sylvatica*) and oak (*Quercus sp.*). There are regional differences in forest damage, and if trees are not severely affected the damage is reversible.

A range of damage types are now recognised. However, the simultaneous appearance of these different types of damage within a short space of time over almost the whole of West Germany indicates a common, as yet unknown, factor. There are several current hypotheses concerning the causes of this damage. The first involves *multiple stress* in which air pollution impairs plant metabolism and makes the plant more susceptible to other stresses such as climatic factors or nutrient deficiency. This hypothesis is difficult to test. *Soil acidity* may be a contributory cause in which aluminium impairs the tree root system. However, damage has been observed on trees growing on a range of soils including calcareous soils in the Alps. *Ozone* may interact with acid mist to induce or exacerbate nutrient deficiences.

Magnesium has been implicated in the high altitude spruce damage; magnesium concentrations in these soils are marginal for tree growth and a series of dry years may have reduced the magnesium supply to the trees below a critical level. Finally, in the Netherlands excess *ammonia deposition* in the order of $5\,\mathrm{g\,N\,m^{-2}\,yr^{-1}}$, derived mainly from animal slurries, has contributed to an increase in the susceptibility of pines to climatic and biotic stresses, mainly due to the acidity produced from NH_4^+ when it enters the soil (section 4.3). It is unlikely, therefore, that a single mechanism is responsible for the range of types of forest damage observed.

The estimates for inputs of acidity into UK soils (Table 4.1) show that wet and dry deposition together may account for as much acidification as that produced by vegetation (section 5.4.3) and nitrification (section 2.4.2).

4.4 Heavy metals

Heavy metals may be introduced into soils from fallout from smelters and the dumping of smelter and mine wastes, urban and traffic sources, and from the application of contaminated *sewage sludge* (particularly in industrial areas). The extent of this pollution can be considerable, for example, approximately 5×10^5 t (dry weight basis) of sewage sludge are applied to land in the UK each year and this may contain high concentrations of copper, zinc, nickel, cadmium, lead and chromium, derived from industrial sources. These heavy metals are strongly adsorbed to soil components and accumulate in soil, remaining possibly for thousands of years.

Although many heavy metals serve as micronutrients for micro-organisms, all are toxic in high concentrations due to their ability to denature proteins. The effects on soil micro-organisms are difficult to quantify due to effects of soil minerals, organic matter and dissolved ions on the solubility and reactivity of the metals. Furthermore, micro-organisms may decompose polluting organic materials and contaminated litter to release metals.

Resistance to heavy metals may develop in soil bacterial populations, but this may take several years. There is some evidence that bacteria isolated from heavily polluted soils are more resistant to heavy metals than those isolated from less polluted soils. *Metal resistance* in bacteria is usually determined by genes carried on plasmids (section 6.4).

The effects of heavy metals on soil biological processes may persist for decades after the input of pollutant has ceased. In experiments at Rothamsted, soils were treated with sewage sludge (organic matter plus heavy metals) or farmyard manure (organic matter but no heavy metals) for 20 years. This resulted in a threefold increase in the amounts of zinc, copper, nickel and cadmium in the soil treated with sludge. Growth of grass was unaffected, but the microbial biomass was reduced by 50%, free-living nitrogen fixation was reduced by 90% and clover rhizobia were rendered ineffective even 20 years after applications of sewage sludge had ceased.

In woodland polluted with heavy metals, plant litter accumulates mainly due to a decrease in the activity of soil invertebrates. Measurements of concentrations of heavy metals in soil or vegetation may not be reliable indicators of the concentrations of metals in primary consumers. Invertebrate animals such as earthworms, slugs and snails, and woodlice

Figure 4.3 The spider *Dysdera crocata* attacking a woodlouse (*Porcellio scaber*). (Courtesy of Dr S.P. Hopkin, University of Reading.)

accumulate heavy metals, and woodlice are particularly useful indicators of the biological availability of metals in polluted soils. The accumulation of up to 1% copper and zinc in woodlice is likely to present a major deterrent to predators such as the spider *Dysdera sp.* which feeds almost exclusively on this soil animal (Figure 4.3).

A number of species occur naturally on soils containing heavy metals. The widespread grass *Agrostis capillaris* grows on metalliferous-mine workings and has evolved races tolerant of copper, cadmium, nickel, lead and zinc, although tolerance of these metals is not apparently linked by a common mechanism. Leadwort (*Minuartia verna*) is only found on calcareous lead spoils. Few plants can exclude metal completely, and some accumulate very high concentrations. The tropical tree *Sebertia acuminata* which grows on ultramafic (nickel-rich) soils in New Caledonia contains up to 11% nickel (wet weight basis) in its blue green latex. Ericoid mycorrhizas (section 3.7) may increase the tolerance of plants to heavy metals by reducing metal uptake; for example, *Calluna vulgaris* is able to colonise copper mine spoil.

4.5 Radioactivity

Soils may be exposed to radioactivity due to leakage from nuclear installations and from buried radioactive waste. High-level radioactive wastes and transuranic wastes, mainly in the form of spent fuel from reactors and fuel reprocessing and from plutonium production for nuclear weapons, is disposed of by long-term isolation from the environment. Low-level wastes, which are generated by all activities involving radioactive materials, are disposed of by incineration or burial in shallow trenches up to 6 m deep. Soil may also be treated with γ-radiation for experimental purposes.

Many soil insects are killed by relatively low doses of radioactivity (0.001–0.05 Mrad) with the most active animals showing the greatest sensitivity. The sensitivity of cells to radiation generally increases with increased metabolic activity. For example, vegetative cells are more sensitive than spores and cysts. Fungi are the most sensitive group of micro-organisms. Viability is decreased by radiation doses as low as 0.0025 Mrad and *Penicillium sp.* and *Trichoderma sp.* are particularly sensitive. 1 Mrad eliminates actinomycetes and 2.5 Mrad sterilises most soils completely. The activity of many soil enzymes shows little decrease after a radiation dose sufficient to inactivate all micro-organisms, and this technique is useful in studying extracellular enzyme activity in soils (section 2.5).

Many micro-organisms are readily able to colonise soil following irradiation, although nitrification activity is not always re-established. The lysis of cells during and after irradiation leads to an increase in available nutrients such as nitrogen, which often results in increased plant growth. There may also be decreases in plant growth, but the reasons for this are not known. The effects of long-term exposure of soils to γ-radiation have received little attention compared to studies of the effects of acute treatments.

When considering the effects of pollution of soils with radionuclides, it is important to distinguish between the effects of the radionuclide itself and the associated radiation. For example, soil micro-organisms are generally tolerant of plutonium, but its toxic effect is due to radiation. Leachate samples from disposal sites for low-level waste indicate that significant microbial activity occurs at these sites. Micro-organisms may solubilise radionuclides such as insoluble plutonium dioxide by the production of chelates. The accumulation of ^{60}Co and ^{137}Cs by micro-organisms followed by movement and eventual death of cells can lead to movement of radionuclides through soils. Our understanding of the ways in which radiation disturbs metabolic processes and upsets the balance between organisms in soil is far from complete.

4.6 Soil moisture

Soil comprises individual particles which vary greatly in size (section 1.2.1) and between the solid particles are pores of a wide range of sizes filled with either aqueous solution or air. The sizes of the solids and spaces have a profound effect on the soil water status. The *potential* of the soil water (a reflection of the work needed to remove unit quantity, and hence its availability) is of more direct importance for soil organisms than is the absolute *water content.*

The soil matrix is very complex and gives rise to forces associated with the interfaces between water and solids, and between water and air. These forces decrease the potential of the soil water and this component of potential is called the *matric potential.* As the soil dries the larger pores (strictly the pores with the largest necks) are emptied first (Figure 4.4). If it is assumed that soil pores act as capillary tubes, then the pore size can be related to the matric potential of the water contained within that pore.

Field capacity is the water content of the soil draining from saturation when the loss due to gravity has become negligible. It corresponds to a matric potential of about -10 kPa (equivalent to -0.1 bar or -100 cm water column). *Permanent wilting point* is the water content at which plants permanently wilt, and is taken as the water content corresponding to a matric potential of -1500 kPa. These criteria have been used to classify the different types of pores found in soils according to size and function. *Transmission pores* ($> 50\ \mu$m) are drained at field capacity, but allow root penetration. *Storage pores* (50–$0.5\ \mu$m) hold water against drainage between the limits of field capacity and permanent wilting point, this water being available for transpiration. *Residual pores* ($< 0.5\ \mu$m) hold water that is unavailable to plants, but which may be evaporated from the soil surface.

The values of water potential at both field capacity and permanent wilting point are to some extent arbitrary, and do not have any precise biological significance. However, well-drained soils are rarely wetter than field capacity, and surface soils may be much drier than permanent wilting point even though plants show no signs of stress, because they are supplied with water from deeper in the soil. At matric potentials less than -1500 kPa movement of water as vapour becomes increasingly important. The relationship between pore size, water potential and the size of various organisms is discussed in section 4.7.

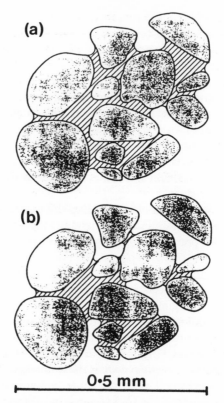

Figure 4.4 Schematic diagram of a section through soil showing the distribution of water at (a) − 10 kPa and (b) − 20 kPa. (From D.M. Griffin (1972) *Ecology of Soil Fungi*, Chapman and Hall, London.)

Water moves in soil from areas of high water potential to areas of low water potential. Differences in water potential between the root and soil, and the *resistance* to movement of water determine the ability of a root to take up water for *transpiration*. The resistance to flow depends upon the conductivity of the soil to water (the *hydraulic conductivity*), which varies greatly with soil water content. The hydraulic conductivity of the soil has little direct influence on micro-organisms, in contrast to plants (section 4.10.1). However, hydraulic conductivity decreases rapidly as water content decreases, and this can affect movement of organisms as they are carried along with the soil water, and slows down the redistribution of water in dry soils.

The potential of the soil water is also decreased by the presence of solutes, which are necessary for growth, and this gives rise to an *osmotic* component of potential (section 4.2). Furthermore, as soils dry the pathways through which solutes can diffuse from areas of high concentration to low concentration also decrease. Therefore, effects of moisture stress on micro-organisms in soil are confounded by several factors (see also section 4.7).

Bacteria and fungi are affected similarly by decreases in soil water potential. Growth is generally restricted to values greater than between − 6000 and − 8000 kPa. At water potentials below − 14 500 kPa *Aspergillus sp.* and *Penicillium sp.* predominate. There is no evidence of a specific xerophytic population in arid zones; fungal species capable of growth in laboratory media at − 30 000 kPa are as prevalent in soils from an English pasture and an Australian rain forest as in desert soils. The apparent dominance of streptomycetes in many dry soils, as determined by plate counts, is probably due to the resistance of their spores to desiccation rather than to hyphal activity at low water potentials.

4.7 Movement of organisms in soil

Bacteria are not uniformly distributed through soil, but are found in discrete colonies covering only about 0.1% of the available surface area. It seems likely that bacteria, fungi and their predators need to move through soil to seek substrates or prey, and to avoid unfavourable conditions such as acidity. Growth of plant roots will assist both in movement of substrates to micro-organisms and in movement of micro-organisms attached to the root surface. Cultivation of soil will also assist in bringing substrates and organisms together. Micro-organisms may also be transported in the guts of soil invertebrates (section 1.3.2.7).

Most bacteria move by rotating semi-rigid protein flagella, approximately 20 nm thick. In *Escherichia coli* the flagella are arranged randomly over the cell surface (*peritrichous flagella*) whereas in *Pseudomonas aeruginosa* there are one or two flagella in a polar position (*polar flagella*). Both types of flagella produce forward motion when rotated, generally in an anticlockwise direction. Reversal of direction of rotation by polar flagella reverses the direction of movement, but with peritrichous flagella this causes deformation of the flagella which results in a tumbling motion. In a uniform environment bacteria with polar flagella spend equal amounts of time travelling in each direction, and rely upon Brownian collisions at the instant of reversal of direction of rotation of flagella to change course. Bacteria with peritrichous flagella perform a three-dimensional random walk during which they swim in a straight line then tumble for approximately 0.1 s in every 1.0 s. Polar flagella (5 μm long) are shorter than peritrichous flagella (15 μm long) and result in faster speeds. For example, *Pseudomonas sp.* can travel at 70 μm s^{-1}.

In a competitive environment such as soil the ability to move in a favourable direction is likely to provide an advantage to an organism. Many bacteria associated with plants show *chemotactic responses* (move towards a specific chemical) in the laboratory, but the significance of this in soil is often unclear. Many obligate or facultative anaerobic bacteria migrate preferentially towards an environment with a higher oxygen concentration. Some bacteria isolated from muds are able to orient themselves and migrate towards the North or South magnetic poles. Root-nodule bacteria are chemotactic to many compounds released from legume roots such as amino acids, sugars and glycoproteins. Chemotaxis does not appear essential for nodulation (section 3.4.2), but may provide a competitive advantage. Motile rhizobia can move at 2 cm day^{-1} through soil and this may increase the chances of the organism meeting a root. The role of chemotaxis in the biology of soil protozoa has not been studied.

For appreciable movement of bacteria through soils there must be enough water-filled pores of the required diameter to provide a continuous pathway. For a rod-shaped bacterium such as *Pseudomonas aeruginosa* pore necks of radius less than 1 μm (corresponding to a matric potential of -300 kPa) are likely to restrict the rate of movement. For *P. aeruginosa* it is estimated that an additional 0.11 cm^3 water cm^{-3} soil is required to provide the necessary continuous pathway. The critical water potential for movement will therefore depend upon the soil texture, and movement of bacteria will generally be restricted to water potentials greater than between -5 and -500 kPa. At lower potentials fungi and actinomycetes assume a greater role, because in the absence of a continuous water pathway of the necessary dimensions hyphae will remain able to spread along the walls of drained pores or across pores from one side to another.

Protozoa and zoospores require pores larger than 5–10 μm for movement and these pores will drain at between -20 and -30 kPa. Such pores will only be filled with water in very wet soils, or during and after heavy rain. The movement of these organisms in the mass flow of water through cracks and channels formed by earthworms and dead roots (*macropores*) may be more important than movement through the soil matrix. There is evidence that coliform bacteria introduced with sewage sludge and animal slurry also move preferentially through macropores. This pathway of movement by micro-organisms is likely to be important in determining the fate of genetically-engineered micro-organisms in soil (section 6.4).

Nematodes move along solid surfaces by undulatory movement in which waves pass along the body from head to tail. Nematodes require pores 30–100 μm in diameter; these pores are drained at field capacity, and movement is through films of water covering the surface of soil particles (section 4.9). When the water films become about 1 μm thick the surface tension forces hold the nematodes too tightly against the soil particle to allow movement. Moisture stress is rarely a problem because movement ceases at water potentials much higher than that needed to desiccate a nematode. Similarly, osmotic stress is seldom a problem. Movement is inhibited in saturated soils due to lack of oxygen.

4.8 Soil atmosphere

Biological responses to changes in soil water content may not be directly related to water but may be due to associated changes in the soil atmosphere. The soil atmosphere differs from the free atmosphere because plant roots and organisms living in soil remove oxygen and produce carbon dioxide. This causes diffusion of oxygen into the soil and carbon dioxide out of the soil. The net result depends upon the relative rates of these different processes. The rate of respiration in soil depends upon soil moisture content (section 4.6), temperature (section 4.1) and amount of readily decomposable organic matter. Measurements made on a soil at Rothamsted Experimental Station indicated rates of oxygen consumption of $0.7–24.0\,\mathrm{g\,m^{-2}\,day^{-1}}$ and rates of carbon dioxide production of $1.2–35.0\,\mathrm{g\,m^{-2}\,day^{-1}}$. Soil contains only enough oxygen to support respiration for about two days if the supply of oxygen is restricted.

Oxygen is unlikely to limit biological activity at low matric potentials because most pores are filled with air in which the diffusion rate of oxygen is 10^4 times as great as that in water. However, at higher water potentials many of the pores are filled with water, and oxygen may become limiting. In a well-structured soil the individual particles are aggregated into crumbs containing very small pores, therefore the crumbs may remain saturated with water over a wide range of water potenials. Larger pores exist between crumbs and they will be more susceptible to draining at relatively low water potentials.

If micro-organisms are evenly distributed within a soil crumb their respiratory activity, coupled with the slow diffusion of oxygen through the crumb, can give rise to anaerobic zones at the centre of water-saturated crumbs of diameter greater than about 6 mm. Such an anaerobic microsite

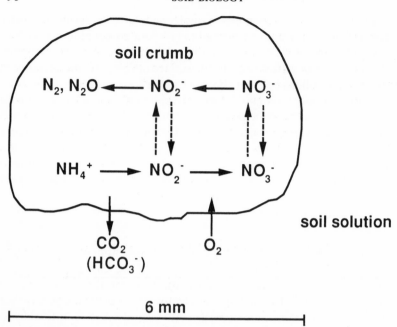

Figure 4.5 Schematic diagram of spatial location and interactions of nitrogen transformations within a soil crumb.

is likely to support denitrification, whereas the more aerobic outer surface of the crumb may be the site of nitrification. Both of these processes may therefore occur simultaneously at sites separated by only a few millimetres, and the reaction intermediates and end products may diffuse through the crumb (Figure 4.5).

The diffusion of volatile compounds such as ethylene, a plant growth hormone (section 3.2.3), will also be affected by the soil water content, but little is known of this.

4.9 Role of surfaces

There is considerable evidence that some solid soil components, in particular some types of clay minerals, affect microbial activity. These effects appear to be mainly indirect rather than involving direct interactions between soil solids and micro-organisms. For example, the stimulatory effects of montmorillonite on fungi and actinomycetes, and the ability of clays to protect micro-organisms against the inhibitory effects of osmotic potential are mainly due to the ability of clay minerals to modify the local physical and chemical conditions in the soil. However, micro-organisms are not readily leached from soil and are often difficult to dislodge, which indicates that they do adhere to soil components.

If micro-organisms do attach themselves to solid surfaces in soil, it is not known how many layers of cells are formed. Surfaces in soil will not be homogeneous; although most of the surface will be clay minerals, there will also be coatings of hydrous metal oxides and organic matter. It may be that clay minerals concentrate organic substrates at their surfaces allowing micro-organisms to grow in what would otherwise be unfavourable conditions. However, if clay surfaces are enriched with nutrients it is not clear whether these nutrients are available for microbial growth. Clay minerals may alleviate the toxicity of heavy metals (section 4.4) to soil micro-organisms by increasing the cation exchange capacity of the soil, and increase the pH buffering capacity of the soil. They may also bind antimicrobial compounds which might otherwise be detrimental to micro-organisms.

Clays offer some degree of protection to micro-organisms against desiccation. The surfaces of these minerals retain water which, together with the coordination water from exchangeable cations, may assist microbes to survive dry conditions. However, this water is unlikely to be available to micro-organisms and the surface tension forces in such films of water may prevent microbial movement. It has been suggested that the higher affinity of montmorillonite than cells for water causes rapid drying of rhizobia, which increases the tolerance of these bacteria to desiccation. Overall, the mechanisms by which surfaces influence microbial survival and activity in soils are poorly understood.

4.10 Effects of roots on soil properties

Roots are important sources of substrates for soil organisms, and the rhizosphere (section 3.1) provides a localised environment in which microbial activity is increased. However, soil physical and chemical conditions in the rhizosphere may also be quite different from conditions in non-rhizophere soil. In addition to the indirect effects of roots on soil properties, for example, oxygen consumption and polysaccharide production by micro-organisms utilising root-derived material, roots may directly affect the rhizosphere soil.

4.10.1   *Physical effects of roots*

Roots generally follow pores and channels that are not much less in diameter than their own; in smaller channels they do not grow freely unless some soil is displaced as the root advances. For example, a pea (*Pisum sativum*) root penetrating clay compacts the soil and orients clay particles within 1 mm of the root surface. In larger channels, such as those formed by earthworms, poor hydraulic contact may exist between the root and soil restricting the uptake of water. This effect of lack of contact may be reduced by the proliferation of root hairs in the humid air of large soil pores. Even though the pores of diameter greater than 10 μm are drained at -30 kPa a thin film of water is always present on the root and root hair surfaces. Such a film of water could be important in allowing movement and growth of micro-organisms in otherwise dry soil.

The *hydrophilic* mucilages (section 3.1) secreted by roots, which are attractive to water, form a layer 1–5 μm thick which assists in the formation of a bridge across which ions and neutral solutes may pass. This could allow plants to take up water which would otherwise be unavailable in dry soils. Mycorrhizal hyphae may also assist in maintaining contact between roots and soil. Such mechanisms will be particularly important as the soil dries out, for instance wheat roots can shrink by 60% of their original diameter at -1000 kPa. Surface-active materials may be produced by roots or the rhizosphere microflora which can change the surface tension and hence the water potential at a particular water content.

Simulation models predict a lower water content around the root compared to the bulk soil, particularly in a dry soil (hence a low hydraulic conductivity, section 4.6) with a high demand for water by the root. However, such predictions depend greatly upon the water uptake per unit root length and this is extremely difficult to measure. The most rapid uptake of water occurs 10–100 mm behind the root tip in solution culture. The interfacial resistance (and hence gradient in water potential) between root and soil may not be important until the soil moisture content is less than 20% of its saturation value.

There may be situations in dry soils in which the water potential of the rhizosphere soil is lower than that of the root, and water therefore moves from the root into the soil. This could happen with a deep-rooting plant in a dry soil at night time when stomata are closed, and the water potential of the soil around roots in the surface layers is lower than the water potential of the soil around deeper roots. In such circumstances, the roots would provide a relatively low resistance pathway for water movement from wet to dry soil. There is no evidence that the resistance to flow of water from roots is different from the resistance of flow of water into roots.

A similar mechanism has been proposed for the desert tree *Prosopis tamarugo* which thrives in the deserts of northern Chile in which the groundwater is often over 40 m deep and no contact occurs with the root system. When the relative humidity is high the plant absorbs water through its leaves and this water then flows down to the root system and into the rhizosphere where it is stored and re-absorbed as required. Such changes in soil water content around roots could have important consequences for the uptake of nutrients such as phosphate, for local growth and activity of roots, and for the activity of the rhizosphere microflora in dry soils.

4.10.2 Chemical effects of roots

Electrical neutrality will be maintained across the root surface in response to the uptake of cations and anions. Analysis of plants indicates that if nitrogen is absorbed as NO_3^- then plants generally absorb more anions than cations. This should lead to an outflow of anions, commonly HCO_3^-. If plants absorb nitrogen as NH_4^+ or as nitrogen gas (nitrogen fixation) then the cation:anion balance indicates that cations, usually H^+, will move out. This is confirmed in experiments in which plants utilising NO_3^- cause an increase in the alkalinity of the rhizosphere soil, whereas plants utilising NH_4^+ or gaseous nitrogen cause localised acidification. Differences of up to one pH unit have been reported between rhizosphere and non-rhizosphere soil. The pH change at the root surface will depend upon the rate of release of HCO_3^- and H^+, the soil pH buffer capacity and other factors. Such changes in acidity may have important consequences for micro-organisms colonising soil around roots, and the manipulation of rhizosphere pH has been proposed as a method for controlling root pathogens such as *Gaeumannomyces graminis*.

The rhizosphere of rape (*Brassica napus*) has been studied in detail because of the plant's ability to take up phosphate from phosphate-deficient soils. This appears to be due to the acidification of the rhizosphere (by up to 2.5 pH units) and increased solubilisation of phosphorus by the plant in response to reduced uptake of NO_3^- under conditions of low phosphate. Different nutrients may be taken up by different parts of the root; this may cause the rhizosphere pH value to vary along the root.

In addition to the outflow of bicarbonate ions, carbon dioxide is liberated at the root surface by respiration. This is unlikely to cause a local acidifying effect because under aerobic conditions carbon dioxide diffuses rapidly

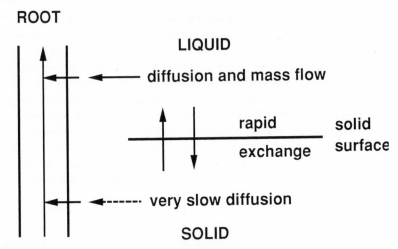

Figure 4.6 Solute transport processes near an absorbing root. (After P.H. Nye and P.B. Tinker (1977).)

away from the root through the air-filled pore space. In contrast, the bicarbonate ion is confined to the solution where its mobility is 10^4 times lower than gaseous carbon dioxide. In calcareous soils the effect of carbon dioxide may be significant.

The concentration of other ions in the rhizosphere may be affected by plant uptake and supply from the surrounding soil. The major transport processes occurring near the root surface are shown in Figure 4.6. Rapid equilibrium takes place between solutes in the soil solution and those adsorbed on the adjacent solid surfaces. These adsorbed solutes tend to *buffer* the soil solution against changes in concentrations induced by root uptake. It is not clear whether the rate of absorption of solutes by roots is determined by their concentration in soil solution or by their concentration within the plant.

Solutes move to the root by *mass flow* (movement of the soil solution caused by transpiration) and by *diffusion* (movement of solutes in response to a concentration gradient), with both processes operating simultaneously. Calculations indicate that more than sufficient calcium and magnesium (except in sodic or very acid soils) and sodium to satisfy the plant demand is transported to the root surface by mass flow, but insufficient potassium, and especially phosphorus.

If a solute is absorbed at a fast rate relative to water, for example phosphorus and potassium, then the solution concentration at the root surface decreases (Figure 3.7). Some of this concentration change may be buffered by the release of ions from solid surfaces, but the remaining concentration gradient will lead to diffusion of ions towards the root. If water is absorbed at a fast rate relative to the solute then the solute concentration at the root surface increases, leading to a localised increase in salinity and a decrease in osmotic potential. These effects may be reduced by the diffusion of solutes away from the root. Therefore, the processes of mass flow and diffusion can lead to zones of nutrient depletion or accumulation around the root (Figure 4.6 and Figure 3.7). The low concentration of phosphate in the rhizosphere (< 0.1 μM) may have important consequences for colonisation of the root region and micro-organisms with reserves of polyphosphate may be favoured.

CHAPTER FIVE

BIOLOGICAL PROCESSES, SOIL FORMATION AND DEVELOPMENT

In Chapter 4 soil was considered as an environment for organisms. However, organisms can alter their environment and cause changes in soil properties. These range from the weathering of bare rock by micro-organisms to changes in a soil profile (section 1.2) brought about by a change in vegetation. This chapter describes the role of soil organisms in the formation and development of soils (*pedogenesis*).

5.1 Pedogenesis

Soil can be described as a three-dimensional body within a landscape, comprising inorganic and organic material which changes with time. It is formed from weathered rock and mineral material, but the point at which this material becomes soil is not clearly defined.

A variety of factors contribute to the formation of a soil profile. A *parent material* of low base status produces an acid soil, whereas a material of higher base status gives rise to more alkaline soils. The shape of the land surface (*topography*) is important; for example, waterlogging in low-lying, poorly-drained areas causes localised reduced zones indicated by gleying or mottling of the soil (section 5.4.1). *Climate* can affect the rate and extent of leaching of bases, this being greater in wet tropical climates. *Biological factors* such as the type of vegetation are also important, for example, conifer trees produce acid leaf leachates which may mobilise some elements in soil and give rise to distinct eluviated horizons, as seen in podzolised soils.

A particular soil is therefore the net result of biological, topographical and climatic factors interacting with a particular parent material; the balance between these factors may alter with *time*. The various processes contributing to soil genesis can be considered as *inputs, outputs, transfers* and *transformations* (Figure 5.1). The resulting profile is the net effect of

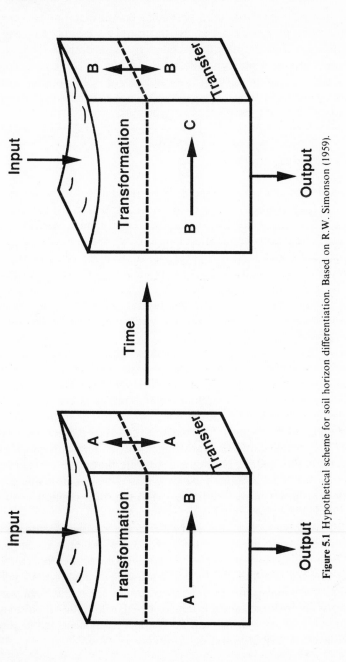

Figure 5.1 Hypothetical scheme for soil horizon differentiation. Based on R.W. Simonson (1959).

these processes operating on a particular parent material in a particular climate. These four processes can be illustrated by reference to the nitrogen cycle (section 2.4): inputs of nitrogen are often derived from nitrogen fixation, outputs from denitrification, transfer of organic nitrogen is mediated by soil animals, and transformation of NH_4^+ to NO_3^- is carried out by nitrifying bacteria. The activities of man may influence pedogenesis by affecting any of these factors or processes.

5.2 Weathering of rock

The weathering of rock is one of the prerequisites for soil formation. Exposed rock is often colonised by micro-organisms, and even in the Antarctic dry valleys, possibly the harshest environment on earth, rocks are colonised by cyanobacteria. Microbial activity contributes to the chemical, physical and mineralogical changes which are involved in weathering. A range of processes such as acid hydrolysis, complex formation, oxidation and reduction, and exchange reactions cause solubilisation of minerals. Some elements may also become less soluble, for example by oxidation of iron and manganese, by reduction of sulphur compounds and by formation of carbonates.

The uptake of ions such as K^+ by micro-organisms and plants can lead to the exchange of ions into solution from minerals. For example, in laboratory culture the fungus *Aspergillus fumigatus* can use micas as a source of potassium; as potassium is immobilised by the fungus, Na^+ in solution exchanges for K^+ in the micas which results in the formation of vermiculite. Wheat plants (*Triticum aestivum*) and seedlings of a range of coniferous and deciduous trees growing in sand culture with potassium and magnesium supplied as biotite, can convert biotite to vermiculite and kaolinite within 12 months. However, it is not clear from these studies whether the observed weathering is due to the roots themselves or to the associated rhizosphere microflora (section 3.1).

Physical forces such as those involved in the freezing of water in cracks and fissures in rocks are important in breaking rocks into smaller particles. Lichens, one of the first organisms to colonise exposed rock, may also play a part in this process. A thallus or hypha growing through a crack will exert pressure which may cause further disintegration. It has also been suggested that the gelatinous substances produced by this symbiotic organism when it dries on the rock surface may tear away mineral fragments in the same way that a layer of drying gelatin does on glass (Figure 5.2).

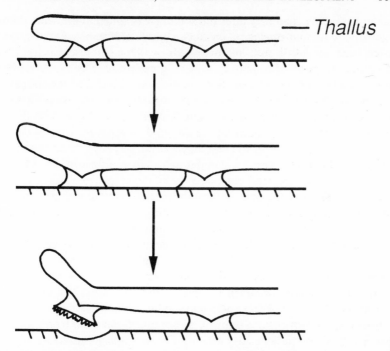

Figure 5.2 Schematic diagram of the breakdown of shale by the lichen *Xanthoria parietina* following drying. (After E.J. Fry (1924) *Annals of Botany, Lond.* **38**, 175–196.)

A decrease in particle size increases the specific area and hence the rate of solubilisation of minerals. There is little *in situ* evidence of a role for organisms in the solubilisation of minerals, but many laboratory experiments have shown that micro-organisms can solubilise a range of elements and transform minerals. *Thiobacillus sp.* have been isolated from particular types of lesions in limestone. The characteristic scaling of limestone buildings (swelling, cracking and eventually falling away of the surface layer to leave a powdery zone) is linked to the conversion of calcium carbonate to calcium sulphate by these bacteria. Basic ferric sulphates are also produced by the action of sulphuric acid produced by *Thiobacillus ferrooxidans* on minerals and rocks.

Micro-organisms have the potential to produce a wide variety of complexing or *chelating agents* that can solubilise elements such as aluminium, iron, manganese, nickel, zinc and calcium. Both aliphatic acids, for example oxalic acid and 2-ketogluconic acid, and aromatic acids, for example salicinic acid may alter the solubility of different elements. Many *Clostridium sp.* and *Bacills sp.* produce volatile and semi-volatile organic acids such as acetic and butyric acids which solubilise calcium and potassium from granite sand. Complexing agents such as oxalic acid are more important in the solubilisation of aluminium and iron which may be carried out by *Pseudomonas sp.*. 2-ketogluconic acid has been identified during the microbial weathering of silicates and during the solubilisation of calcium carbonate, montmorillonite and hydroxyapatite. The difficulty is to obtain information on the reactions between specific chelating agents and mineral particles within soil under natural conditions, in which several processes may be operating concurrently. For example, in waterlogged soil the simultaneous production of organic acids by bacteria and the reduction of iron and manganese may lead to the modification of biotite to vermiculite, and illite–vermiculite to montmorillonite.

The colonisation of rock by lichens presents a simpler system in which the environment is more uniform and the antibiotic properties of lichen acids, together with the otherwise unfavourable conditions, limit colonisation by other micro-organisms. Lichens are symbiotic organisms formed by a cyanobacterium or green alga and a fungus, which may fix atmospheric nitrogen and produce large quantities of extracellular acids (comprising as much as 2–5% of the lichen dry weight). The rock lichen *Parmelia conspersa* produces salacinic acid which when added to ground granite and mica reacts within a few hours. The intimate contact between lichen hyphae and rock, together with the production of extracellular acids suggest a major role for these organisms in weathering of rock.

The ability of cyanobacteria and lichens to withstand drought and extreme temperatures, and to fix carbon dioxide and nitrogen, makes them well adapted for primary colonisation of rocks. Morphological features such as a reduced evaporative surface, thickened cortex and various amorphous and sometimes gelatinous external layers, together with pigmentation, assist in the protection of lichens against desiccation and excessive radiation from direct sunlight. Values for cyanobacterial biomass of $2–62\,g\,m^{-2}$ rock have been reported for desert areas in the Middle East and North America.

Apart from a direct effect on the physical and chemical weathering of rocks, these organisms have an indirect effect in providing the first input of organic matter to the weathered rock. Humic acids formed from such organic material are active in weathering processes. For example, iron and aluminium from minerals form soluble complexes with fulvic acid which allows leaching of these elements, as may occur during *podzolisation*. The reducing properties of fulvic acid and polyphenols may also increase the mobility of iron as it is converted from Fe^{3+} to Fe^{2+}, and this may allow podzolisation to occur under aerobic conditions. The organic matter produced by the primary colonisers allows growth of heterotrophic organisms (section 2.3.1) and subsequent colonisation by plants. The nitrogen fixed by lichens may be readily available to plants; experiments using ^{15}N-labelled nitrogen showed that part of the atmospheric nitrogen fixed by lichen crusts in desert grassland soils became available to plants growing in the crusts within 160 days. These later colonisers of rock and rock-derived material may then cause further weathering.

5.3 The role of organic matter

As weathered rock accumulates organic matter it begins to show features of soil. Organic matter affects soil fertility, structure and profile development. These effects are often interrelated, for example, the fertility of the soil will influence the types of plants which grow, and the type of plant may influence the development of the soil profile. Soil organic matter is intimately linked to the microbial biomass (section 1.3.1), and soil organisms play a major role in the turnover of this material.

The accumulation of organic matter leads to horizon differentiation in most soils. The effect of organic matter on soil processes will depend not only upon the amount of organic matter but also on its rate of decomposition. In the early stages of development the input of organic matter is greater than the loss, and as time progresses the gains and losses move towards an equilibrium. In a few relatively rare cases, for example where the vegetation cover has remained the same for a long period of time, it can be assumed that the input of organic matter from the vegetation is constant during that period and steady state conditions exist. Few experiments have tested this assumption. However, if such conditions exist then the rate of decomposition or *turnover* of organic matter can be defined as:

$$\text{turnover time (years)} = \frac{\text{total soil organic matter}}{\text{annual input of organic matter.}}$$

The decomposition of organic matter is a complex and poorly understood process which involves soil animals and micro-organisms (section 1.4). It is influenced by the nature or quality of the organic matter (section 2.2), the climate, the soil conditions such as acidity (section 4.3), and especially man's actions (section 5.5).

Table 5.1 The turnover of organic carbon in soils from different ecosystems (after Jenkinson, 1981).

Ecosystem	Net primary production $(\mathrm{g\,C\,m^{-2}\,yr^{-1}})$	Carbon input to soil $(\mathrm{g\,m^{-2}\,yr^{-1}})$	Carbon in soil $(\mathrm{g\,m^{-2}})$	Turnover time (yr)
Temperate arable	260	120	2600	22
Temperate forest	710	240	7200	30
Subhumid savannah	140	50	1700	34
Tropical rain forest	950	490	4400	9

Table 5.1 shows the rates of input and of turnover of organic matter for soils from a range of ecosystems. Only the example of the arable soil is known to be under steady state conditions; this data comes from one of the long-term field experiments at Rothamsted Experimental Station where levels of organic matter have remained constant in soils growing wheat continuously since 1843. Organic matter turnover is fastest in tropical rain forests where the higher temperature and humidity and the action of termites accelerates decomposition (section 5.3.2).

The turnover times given in Table 5.1 are average values for all of the soil organic matter. However organic matter is far from homogeneous (section 2.2.2). Mathematical models have been developed to simulate the turnover of organic matter in soils. These models vary in their complexity, but all of the models partition organic matter into different fractions and then ascribe a turnover time to each particular fraction. Turnover times for these fractions are believed to vary from 1.7 years for biomass to 2400 years for the most resistant fraction.

5.3.1 Role of earthworms

Soil animals play a major role in decomposition of organic matter (the role of micro-organisms is discussed in Chapter 2). Probably the best known example is that of the earthworm (section 1.3.2.7). This was the subject of Darwin's final book *The Formation of Vegetable Mould Through the Action of Worms, with Observations of Their Habits* published in 1881, which marked the culmination of over 40 years' interest in these animals. Darwin's studies led him to the conclusion that 'Worms have played a more important role in the history of the world than most persons would at first suppose'. He observed that worms select their food which they then shred and partially digest and mix with earth to form dark, rich humus which encourages nitrification. He suggested that earthworm burrows allow air to penetrate deeper into the ground and allow the downward passage of roots into soil. Subsequent investigations have confirmed much of Darwin's work.

The formation of *burrows* is a feature of earthworm activity, although not all species have burrows; it is usually only those that go deep into soil, for example *Lumbricus terrestris*. Burrows are formed by worms eating their way through soil and pushing through cracks. They are 3–12 mm in diameter with smooth walls cemented with mucous secretions and ejected soil. The mucous may act as a substrate for growth by fungi. Smaller, shallow-working worms have no permanent burrows.

Burrowing species produce *casts* on the soil surface near burrow exits. There are many different forms of casts, and they are often typical of the species that produced them. European casts are usually < 100 g, but the giant *Notoscolex sp.* found in Burma produces large tower-shaped casts up to 25 cm high and 4 cm in diameter weighing up to 1.6 kg. Darwin estimated the annual production of earthworm casts in English pastures to be 1870–4030 g m^{-2}. More casts are found in pasture soils than in arable soils, and this may reflect differences in organic matter input and earthworm populations. Larger amounts of cast material are reported for tropical soils, for example 5000 g m^{-2} in Ghana. The amount of soil passing through the animals may be larger than these values because some worms void their casts underground.

Earthworms generally prefer soils with near neutral pH values. The absence of worms in acid soils leads to the accumulation of a thick mat of slowly decaying organic matter at the soil surface characteristic of soils with mor humus (section 5.4.3). Soil crumb structure may also be poorer when worms are absent. The importance of earthworms lies in their role in fragmenting plant material, making it more readily decomposed by soil micro-organisms, and in mixing this organic material with the soil. A few common species such as *Lumbricus terrestris* appear responsible for the fragmentation of most of the litter in woodlands of the temperate zone. *Lumbricus terrestris* can remove 90% of the autumn leaf fall in an apple orchard during winter, accounting for 120 g dry matter m^{-2}.

Different species of earthworm have different food preferences, for example *Lumbricus rubellus* is a litter feeder, whereas *Allolobophora caliginosa* consumes partially decomposed organic matter. *Lumbricus terrestris* pulls leaves and other plant material into the mouth of the burrow before feeding on it, thereby plugging the burrow. It carefully selects food material and pulls leaves into the burrow, usually by the tip of the laminae leaving the unpalatable petioles protruding from the burrow.

The ability of most species to distinguish between different kinds of plant litter may be due to differences in plant mineral content or alkaloid content. Weathering of many types of leaves is required before they are palatable to worms; this may allow the leaching of undesirable polyphenolic compounds. Worms prefer moist litter to dry litter and they can consume 100–300 mg plant material g body weight^{-1} day^{-1}. They pass a mixture of organic and inorganic material through their intestines when feeding and burrowing. The intestinal microflora of the worms (section 1.3.2.7) may assist in humification, but the role of earthworms in decomposition as opposed to fragmentation of plant material is not clear.

5.3.2 Role of termites

Termites are often the dominant members of the soil fauna in tropical, subtropical, semi-arid and some warm-temperate regions of the world. A

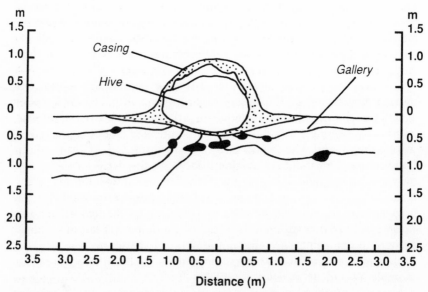

Figure 5.3 The structure of a termite mound formed by *Macroterries nigeriensis*. (After P.H. Nye (1955) *Journal of Soil Science* **6**, 73–83.)

colony of termites forms a characteristic *mound* (Figure 5.3) often when forest has been cleared or thinned for agricultural use, but also in forests, sometimes covering trees and shrubs,

Mounds may be built rapidly, 60 cm in one month, and a single mound may contain 2.5 t earth. 1–10 mounds ha^{-1} have been observed in West Africa. The central part of the mound (the *hive*) has a complex system of *galleries* leading to food sources. Galleries radiate from the floor of the nest to about 90 cm depth and extend for about 3 m on either side of the mound. Galleries open out at regular intervals into chambers up to 90 cm wide. As food supplies are reduced galleries are abandoned and new ones exploited. A colony may last for 10–80 years and the mound is only slowly decomposed after being abandoned.

The colony is virtually a closed system in which unhealthy and dead individuals are consumed, and the only losses are due to predation and annual flights of secondary reproductives (*alates*). The release of alates, which are rich in fat and protein, over 1–2 h at favourable times of the year may represent up to 60% of the colony biomass. Predation on colonies and foraging parties is often by specialised predators, for example in Nigeria parties of up to 300 ants may capture 1000 termites.

Termites feed on all kinds of plant material (wood, grass and roots) both living and decomposing. They are particularly adept at consuming plant material before the material is attacked by saprophytic micro-organisms. However, as most work on organic matter decomposition has been concentrated in cool temperate regions, the role of these organisms has not been fully appreciated.

A few species feed on living plants, preferring dry parts of grasses to fresh green tissue. Dead plant material is often attacked before it falls to the ground. Many termites, for example *Macrotermes sp.*, eat dead material such as bark on living plants. Fresh woody litter is readily consumed. Termites also eat decomposing litter and dung, and there is sometimes a relationship between the stage of decay of the organic matter and the species feeding. This may involve effects of fungi on the material rendering it more readily available or palatable.

Many species consume soil rich in organic matter; these species show a morphological specialisation of the *mandibles* (mouthparts) which distinguish them clearly from those species feeding on decomposing organic matter. There is further specialisation in some species, for example *Hospitaltermes sp.* found in Malaysia feeds on lichens. Practices such as *coprophagy* (eating faeces), *necrophagy* (eating the dead) and *cannibalism* (eating the living) within the termite colony lead to efficient recycling and utilisation of nutrients.

In the savannah, termites influence the initial stages of decomposition but in the rain forest they affect the later stages of decomposition. Most of the energy for termites comes from the breakdown of complex plant polysaccharides. Lignin is degraded and cellulases are produced in the gut, but, as with earthworms, it is not clear whether these enzymes are produced by the animal or by the intestinal microflora. The efficiency of assimilation is 54–93%, and the intense degradation of plant material gives rise to faeces with distinct chemical properties. Faeces have a high C:N ratio (21–107) and, particularly those produced by the wood feeders, have a low carbohydrate content and a high lignin content. Termites are unusual in that they use their faeces for constructing part of the nest system. However, the practice of coprophagy means that excreta may pass through the alimentary system of several individuals before being used.

The Macrotermitinae are a special case because all of their excreta are used to construct fungus combs. These termites are unable to decompose plant structural polysaccharides, but the fungus combs support a pure culture of *Termitomyces sp.* which degrades these compounds, and the products of their activity are ingested by the termites in the old portions of the comb. Combs are replaced by new ones every 5–8 weeks. The termites also eat white fungus spherules, rich in nitrogen, on the surface of the comb. Metabolism of carbon by the fungus leads to a reduction in the C:N ratio of the combs. Termites therefore consume more food than is required for their own maintenance. Also, some Macrotermitinae do not consume all of their food directly, instead they finely macerate the material and store it in heaps to allow conditioning.

An attempt to quantify the role of termites in decomposition is presented in Table 5.2 for savannah woodland in Nigeria (soil-feeders are excluded). The total estimated consumption of $168 \, \mathrm{g\,m^{-2}\,yr^{-1}}$ represents a significant

Table 5.2 Role of termites in decomposition of wood, grass and fresh litter in Southern Guinea Savannah Woodland (after Wood, 1976).

Measurement (dry weight basis)	Macrotermitinae	Others	Total
No. $\mathrm{m^{-2}}$	1928	1540	3468
$\mathrm{g\,m^{-2}}$	1.5	0.9	2.4
Consumption $(\mathrm{g\,m^{-2}\,yr^{-1}})$	129.2	38.3	167.5
Returns $(\mathrm{g\,m^{-2}\,yr^{-1}})$ via			
Faeces	0	12.3	12.3
Alates and neuters	9.0	2.7	11.7

fraction of the estimated grass production of $200-300\,\mathrm{g\,m^{-2}\,yr^{-1}}$ plus litter fall from trees of $200-400\,\mathrm{g\,m^{-2}\,yr^{-1}}$. This rate of consumption assumes a greater significance when the loss of grass, leaf and woody material during bush fires is considered.

5.4 Profile development

The development of a soil profile may be influenced by a variety of biological factors ranging from the development of soil structure by micro-organisms to the effects of vegetation in influencing the extent of leaching of bases from the soil. The activities of some organisms may have an indirect effect on pedogenesis. For example, in the arctic tundra during the long winters lemmings (*Lemmus sibiricus*) live in the grass and sedge layer between the frozen soil and the snow cover. During the brief summer the snow melts and the permafrost thaws allowing plant growth. Grazing of the vegetation cover by the lemmings removes this insulating layer from the soil surface allowing the soil to thaw to a greater depth than if the lemmings were absent, thereby increasing biological and chemical activity in the soil.

5.4.1   Role of micro-organisms

Micro-organisms are responsible for some features characteristic of particular soil profiles. Under anaerobic conditions the reduction of Fe^{3+} (red and insoluble) to Fe^{2+} (dull grey/green and soluble) during periods of waterlogging produces *mottles* characteristic of *gley* soils. It is not clear whether this effect by micro-organisms is indirect via a reduction in oxygen concentration or direct due to the use of Fe^{3+} as an alternative electron acceptor under anaerobic conditions (section 2.3.1). Under acid conditions *Thiobacillus ferrooxidans* produces basic ferric sulphates and weathers minerals that supply K^+ ions to produce the mineral jarosite. These processes contribute to the formation of acid sulphate soils.

Micro-organisms participate in the development of soil structure. The stabilisation of soil aggregates (section 1.2.3), which are formed largely as a result of physical forces, is important for soil development and for crop growth, as it determines the water (section 4.6) and air (section 4.8) relationships for that soil. In most soils, organic binding agents are involved; they may be decomposition products of plants, animals or micro-organisms, microbial cells or products of microbial metabolism.

The addition of organic matter to soil in the presence of micro-organisms improves soil physical conditions. The more readily-available organic matter produces a more rapid response. However, there is no simple relationship between microbial numbers and the extent of soil stabilisation. Polysaccharides appear to be the major factor in aggregate stabilisation, particularly in cultivated soils. Bacteria can produce polysaccharides in laboratory culture and it seems likely that they also do this in soil, perhaps using root-derived material as substrates (section 2.1.1). Bacterial polysaccharides sometimes appear resistant to decomposition in soil. This may be due to the presence of metal cations, or to protection from microbial attack if the polymers are within aggregates.

Fungal polysaccharides are of lower molecular weight than bacterial polysaccharides, and a decrease in the molecular weight of the polymer is often associated with a decrease in adsorption energy. Polysaccharides extracted and fractionated from soils comprise a complex mixture of sugar units. The binding activity of polysaccharides is due to the length and linear structure of the molecule which allows it to bridge spaces between soil particles. Their flexibility allows many points of contact so that van der Waal's forces can be more effective. Also, the hydroxyl groups allow hydrogen bond formation, and the acid groups allow ionic binding to soil particles via multivalent ions (cation bridges).

Bacteria are readily adsorbed to the surface of soil particles (section 4.9). The functional groups present on the bacterial cell surface probably allow a mechanism of binding similar to those for organic polymers described above. Bacteria surrounded by clay crystals could form a microaggregate whereas fungi are probably responsible for binding larger soil particles. Fungal hyphae retain their strength after they die allowing a more permanent effect. Fungi may also form aggregates by bringing soil particles together as they grow. Figure 5.4 represents the possible effects of microorganisms on aggregate stability.

While smaller aggregates ($2-20\,\mu$m) are probably bound together by organic bonds, larger aggregates ($>2000\,\mu$m) may be held together by a network of roots and fungal hyphae. Plant roots increase the stability of the surrounding aggregates and this effect may be partly due to mycorrhizal hyphae (section 3.7). It has been observed that lucerne (*Medicago sativa*) when grown as a break crop improves the structure of soil.

5.4.2 Role of soil animals

The burrowing activity of some soil animals has important consequences for pedogenesis. The term *crotovina* was introduced by Russian workers to

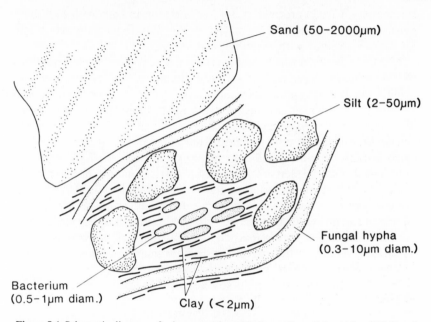

Figure 5.4 Schematic diagram of micro-organisms binding soil particles. (After J.M. Lynch and E. Bragg (1985).)

describe the burrows found in the black chernozem soils of the Russian grain belts. These are formed by the inwash of surface soil into the burrows of hibernating animals such as marmots.

A cicada crotovina, a cicada burrow filled with material from the horizon in which the burrow occurs, has been described for certain soils in the western USA where there is extensive burrowing by these animals. Newly-hatched nymphs of species such as *Platypedia sp.* drop to the ground, burrow into the soil, attach themselves to plant roots and feed on plant sap. Their life-cycle lasts for 2–6 years. The nymphs are most numerous at depths of between 30 and 100 cm, with the smaller ones deeper in the soil. The nymphs inhabit the open portion of the burrow, up to 75 mm long and 20 mm in diameter. They move through the soil to find food, to escape from low temperatures and to emerge from the soil when mature. They continually backfill the burrows with local material, and up to 75 cm of the profile may be composed mainly of cicada crotovinas. Calcium carbonate may accumulate with time leading to the formation of cemented nodules and cylinders. A cylindrical blocky soil structure is produced as a result of this activity. Cicada crotovinas are less common in soils with high bulk density and a textural B horizon. They prefer well-drained silt loams, and are often associated with shrubby plant species.

A major role for termites and earthworms in profile development has been proposed for a sequence of soils near Ibadan in Nigeria. A sequence of profiles taken across a stream valley developed from granite gneiss and supporting rain forest demonstrate the interrelationship between soils developing on a slope (Figure 5.5). Such a sequence of soils is termed a *catena*. Above the sedentary horizon (S) lies a horizon of soil creep (Cr). This creep horizon is sub-divided into an upper horizon (CrW) comprising humus and material derived from worm casts, a middle horizon (CrT) formed by the action of termites, and a lower horizon (CrG). This CrG horizon comprises quartz gravel concentrated by eluviation of clay into the S horizon and the removal of fine earth fragments into the horizon above by termites. The largest size of particle found in the CrT horizon is 4 mm, which corresponds to the largest size particle that the termites can carry or digest. Worms sort the fine material (maximum size 0.5 mm) from the CrT horizon and deposit it on the soil surface. The worm cast material and

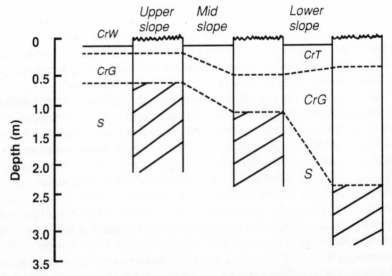

Figure 5.5 Catenary sequence of soils formed near Ibadan, Nigeria, Horizons: *CrW* formed by worms, *CrT* formed by termites, *CrG* gravel accumulation, *S* sedentary horizon. (After P.H. Nye (1955) *Journal of Soil Science* **5**, 7–21.)

topsoil derived from it is estimated to be $30\,kg\,m^{-2}$. Where worms are absent from the soil the topmost layers are formed largely of earth deposited by ants.

The location of the profiles on a slope means that they are in a continuous state of development in which the upper horizons of the soils up-slope are being removed by creep and constantly being renewed by material from the top of the sedentary horizon due to the activity of termites and worms. Soils lower down the slope accumulate material (Figure 5.5). The termites are particularly important here in the formation of a gravel-free topsoil. Also, their activity disturbs the topsoil and accelerates creep causing a continual lowering of the topsoil, allowing the tree roots to maintain close contact with the nutrients released in the subsoil.

5.4.3   Role of vegetation

The complex interactions that occur between soil-forming factors during pedogenesis make it difficult to clearly identify the major causes of profile change. In some cases the dominant factors are more obvious; in the freely-drained soils of north-west Europe, climatic factors cause progressive leaching and consequently increased acidity leading in some soils to podzolisation. However, it has been known for over a century that changes in vegetation are also important. For example, if oak (*Quercus sp.*) is replaced by beech (*Fagus sylvatica*) or heather (*Calluna vulgaris*) the brown forest soil with *mull humus* (neutral pH, high base content and earthworm activity) changes to a podzol soil with *mor humus* (acid, with a mainly fungal population, and with no mixing of organic and mineral soil).

Different plant species and their litter vary in their content of nutrients and secondary chemicals, which in turn can affect the rate of decomposition and may directly affect soil acidity, and accelerate *eluviation* (movement of soil material through the profile). Species also differ in the way in which they modify the composition of rainfall draining off the leaves. Alder (*Alnus sp.*) tolerates acid soils and also acidifies rapidly, with decreases in soil pH values of 0.5–2.0 units recorded in 20–30 years. Similar changes occur under heather (*Calluna vulgaris*) and gorse (*Ulex sp.*). It may take 1000 years for a mature podzol to form, but the early stages of horizon differentiation occur within less than 100 years.

Some species may reduce soil acidity and reverse the trend towards podzolisation. The colonisation of heathland by birch (*Betula sp.*) may increase the soil pH value from 3.8 to 4.9 in 90 years. Bracken (*Pteridium aquilinum*) sometimes reduces acidity and at other times has the opposite effect. Herbaceous species generally do not acidify soils. For example, heavily grazed natural pastures of *Agrostis sp.* and *Festuca sp.* are associated with mull-like humus, partly due to the rapid cycling of litter and nutrients by the grazing animal which prevents organic matter accumulation.

A marked increase in soil acidity and consequent decline in productivity of subterranean clover (*Trifolium subterraneum*), an annual self-regenerating pasture legume, has been observed for a range of soils and farming systems in south-east Australia. Decreases of one pH unit have been recorded over 25–50 years. This is due to a combination of factors (section 4.3), including the accumulation of organic matter giving rise to an increase in cation exchange capacity and exchangeable acidity, and acidification of the legume rhizosphere (section 4.10.2).

The effect of vegetation on soils is evident at a site in Saskatchewan (Canada) where, on uniform parent material with only small differences in climate, grassland is found adjacent to woodland on gently sloping hills. The chernozem soil formed under grassland was changed to a podzol (section 1.2.4) following invasion by trees which caused a decrease in pH, an increase in leaching of calcium carbonate, the appearance of litter and humus horizons, and an increase in clay migration through the profile.

A succession of plant species is often observed in natural ecosystems; species-poor associations of low stature such as grasses and sedges are followed by communities of tall plants with many species and complex structures. A link is often assumed between plant succession and pedogenesis, with natural progression toward a plant climax community on a mature soil. If such a climax community does occur, then if it is undisturbed it may be assumed to be in permanent equilibrium with the environment, to have maximum total biomass, highest species diversity and highest stability. However, if the environment does not remain constant then such a situation will not occur.

A study of five dated river terraces in the Alaskan tundra, all with comparable parent material, climate and topography but formed at different times, provides an example of changes in vegetation and soil with time. The pioneer vegetation (25–30 years old) comprises isolated low shrubs such as *Dryas sp.* and herbaceous plants such as *Astragalus sp.*, both of which are nitrogen fixing plants. On the next older terrace (100 years old) the pioneer shrubs were replaced by willow (*Salix sp.*) and a mat of plants including grasses such as *Poa sp..* A 5 cm deep organic layer of decomposing moss and plant tissue has formed. During the shrub stage (150–300 years old) willow was replaced by birch (*Betula sp.*) and other trees which survive

Table 5.3   Changes in microbial populations and acidity in soil associated with a succession of plant species on sand dunes in Scotland. After D.M. Webley et al. (1952) Journal of Ecology **40**, 168–178.

Vegetation	pH	Bacteria (no. g^{-1})	Fungi (no. g^{-1})
Open sand	6.8	18000	270
Yellow dune	6.7	1630000	1700
Early fixed dune	5.1	1700000	68470
Dune pasture	4.8	2230000	209780
Dune heath	4.3	127000	148000

by putting adventitious roots into the moss layer, which is now up to 35 cm deep. The tundra vegetation on the oldest terrace (5000–9000 years old) comprises low shrub-sedge tussock-moss and cotton grass (*Eriophorum vaginatum*). The plant succession is associated with an increase in the soil nitrogen content.

The succession of plant species on sand dunes is often paralleled by a microbial succession in the soil. On the shores of Lake Michigan (USA) no characteristic microflora was associated with the pioneer species, *Ammophila breviligulata* (marram grass), possibly due to the instability of the sand and the low organic matter content. However, a distinct fungal flora was associated with the later community of *Andropogon scoparius* (bunchgrass).

A succession on sand dunes in Aberdeenshire (Scotland) from *Ammophila arenaria* to *Calluna vulgaris* (heather) was associated with an initial increase in the numbers of bacteria and fungi followed by a decrease in the numbers of bacteria while the fungal population continued to increase (Table 5.3). The reduction in the bacterial population was probably due to an increase in the soil acidity during the later stages of succession. In the early stages of succession, when the soil organic matter content is low, the input of organic material from roots and the rhizosphere microflora (section 3.1) is particularly important.

5.5   The influence of man

Human activity can have a major effect on soil development, and the effect is often rapid. For example, a marsh can be converted into a meadow and a tropical rain forest into an unproductive pasture within a few years. The effect of human settlement is most often via the effect on vegetation, usually associated with the introduction or modification of agricultural practices.

The link between civilisation and soils is illustrated by the rise and fall over a period of 4000 years of the Mesopotamian, Assyrian and Babylonian civilisations in the fertile crescent of the Tigris and Euphrates river basins in present-day Iraq. These fluctuations were associated with increases in crop production due to irrigation followed by a decline due to salinisation (section 4.2) and silting of irrigation canals.

Human activity can alter soil organic matter content, usually as a consequence of changing the vegetation cover. Management practices can alter the annual input of organic matter, for example when crop residues are removed from the soil or when vegetation is burned, and can also alter the rate of turnover of organic matter, for example by the cultivation of soil.

In Ghana, when an evergreen forest was cleared and cropped with a maize (*Zea mays*)/cassava (*Manihot esculenta*) rotation for eight years, the carbon content of the soil was reduced from 2.2% to 1.5%. The cultivation of a virgin prairie soil in Kansas for 40 years reduced the amount of nitrogen in the soil by almost 50%. The introduction of arable farming nearly always leads to a reduction in the soil organic matter content. Conversely, the organic matter content of an arable field at Rothamsted Experimental Station left uncultivated for 90 years increased almost threefold as woodland regenerated.

Modern temperate agricultural methods may be able to sustain production at low soil organic matter levels but this depends upon an input of organic matter from crop residues. In the tropics, however, a fallow period is essential to allow organic matter and nutrient contents of the soil to recover before further cropping (section 6.5).

The natural climax vegetation in the humid temperate zone is broad-leaved woodland associated with a brown earth soil. Evidence obtained from pollen grains preserved in acid soils, and from estimates of mean residence time using ^{14}C-dating of podzol soils in eastern England, indicates a change in soil development caused by the clearance of woodland by Neolithic and Bronze Age man. The heavy canopy of a broad-leaved forest transpires water taken from deep in the soil and also reduces soil evaporation. The removal of this vegetation causes increased drying of the soil surface and a rise in the water table. In Western Australia this is causing an increase in salinity (section 4.2) of soils cleared of the natural forest. Exposure of the soil surface to the atmosphere also causes an increase in the rate of organic matter decomposition. The absence of an input of leaf litter with high base content causes a reduction in the base content of the topsoil and together with the reduced organic matter content this reduces the earthworm population. Podzolisation normally occurs as a consequence, but this may be prevented by cultivation, which brings bases to the soil surface, and by the addition of organic manures and lime.

In recent years large increases have been observed in the area of desert. Much of this land was formerly agricultural land and each year 2×10^5 km^2 deteriorate to the point of zero economic yield. This is due to overcultivation, deforestation, overgrazing and unskilled irrigation. All of these factors, which are exacerbated by increasing population pressure, lead to a decrease in soil organic matter and increased susceptibility to wind and water erosion. The clearing and burning of tropical rain forest, which is rich in calcium and potassium relative to the soil, and its replacement by pasture or plantation forestry often leads to decreases in productivity within a few years. The reduced ground cover and lower inputs of organic matter allow erosion of the soil and further loss of nutrients.

CHAPTER SIX
SOIL BIOLOGY AND MAN

In Chapter 5 the role of human activity in affecting soil development was considered, mainly in terms of the consequences of changes in land use. As our understanding of soil has developed there have been attempts to manipulate organisms or biological processes in soil, either by chemical or biological means, in order to increase plant growth. Almost without exception these attempts to manipulate soil organisms have been made without a complete understanding of the physiology and ecology of the particular organisms. This has led, therefore, to disappointment over the fickle nature of biological methods. At the same time concerns have increased over the side-effects associated with the use of chemicals. This chapter examines some of the ways in which humans have attempted to manipulate the biology of the soil.

6.1 Pesticides

Pesticides are currently an integral part of modern farming practice, contributing to increased agricultural productivity, and are being used increasingly in developing countries where pests and diseases are one of the major factors limiting food production. They include herbicides, insecticides, fungicides and nematicides.

Chemical pesticides, particularly naturally-occurring products such as sulphur compounds, have been used for centuries. However, the use of synthetic organic pesticides originated in the 1940s with the discoveries of the insecticidal action of DDT and of the synthetic plant hormone herbicides MCPA and 2,4-D. The range of products has expanded dramatically over the last 50 years and in 1987 approximately 600 active ingredients were available to farmers around the world. In 1986, 3.5×10^6 t of pesticides were used with a value of almost $\$16 \times 10^9$.

6.1.1   *Side-effects*

Pesticides are by definition biologically active compounds, but it has not yet been possible to achieve absolute specificity, therefore non-target organisms are affected. As little as 1% of the pesticide applied may reach the target, and in the developed world a single crop may receive one or more applications of several pesticides each season. Countries vary in their regulatory requirements concerning the use of pesticides. Pesticide degradation and mobility in soil, together with aquatic toxicology tests, are usually required, and some countries require data on the effects of pesticides on soil micro-organisms.

Pesticides can enter soil by a variety of routes (Figure 6.1). Apart from accidental spillage of chemicals, there may be overspraying or runoff from plants into soil. Faeces of treated animals and death of treated organisms leads to incorporation of compounds into soil. Pesticides may also move from soil to other parts of the environment, for example by aerial transport of soil, plant debris and spores, and by volatilisation, leaching and runoff from soil. Pesticides normally enter the soil at or below the concentrations recommended for agricultural use $(0.05-0.5\,m^{-2})$, which, assuming an even distribution within the top 10 cm of soil, would give a concentration in soil of 0.5–5 ppm.

A compound such as DDT (Figure 6.2a) contains C–C, C–H and C–Cl bonds which are very strong and render the molecule resistant to degradation. Furthermore, it has low solubility in water but is *lipophilic* (dissolves in fat), and therefore accumulates in plants and animals in the food chain. When it was first used, DDT was extremely effective at eradicating the insect vectors of diseases such as malaria and yellow fever, but excessive use led to the development of DDT-resistance in target insects. Many countries have now banned the use of this insecticide.

Insecticides may affect non-target soil animals, but the effects are generally minimal and often transient. Earthworms are not particularly susceptible to insecticides, but they can concentrate organo-chlorine compounds by a factor of up to nine. Different groups of organisms may be affected differently, and this may upset the interaction between predator and prey. For example, DDT causes a decrease in the population of predatory mites which leads to an increase in the population of springtails. However, both mites and springtails (section 1.3.2.8) are susceptible to *nematicides* (used to control nematode diseases). There is little information on the effects of *fungicides* on soil invertebrates.

In temperate cultivated soils the indirect effects of pesticides in allowing

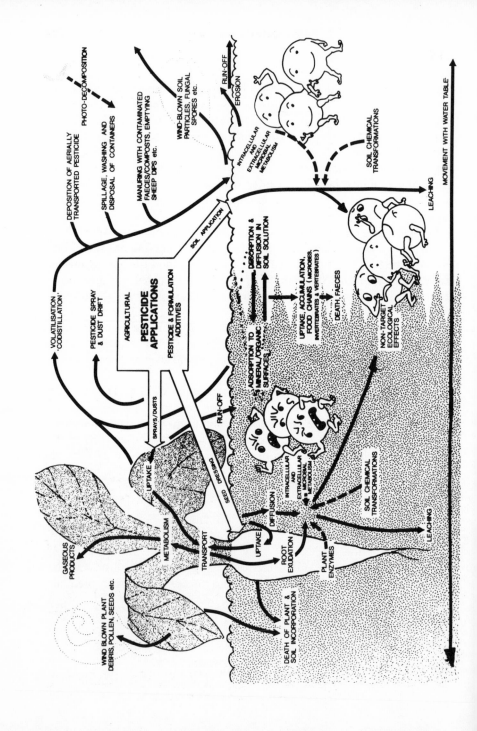

PHOTO-DECOMPOSITION

DEPOSITION OF AERIALLY TRANSPORTED PESTICIDE

SPILLAGE, WASHING AND DISPOSAL OF CONTAINERS

MANURING WITH CONTAMINATED FAECES/COMPOSTS, EMPTYING SHEEP DIPS etc.

WIND-BLOWN SOIL PARTICLES, FUNGAL SPORES etc.

RUN-OFF

EROSION

INTRACELLULAR AND EXTRACELLULAR MICROBIAL METABOLISM

SOIL CHEMICAL TRANSFORMATIONS

LEACHING

MOVEMENT WITH WATER TABLE

VOLATILISATION 'CODISTILLATION'

PESTICIDE SPRAY & DUST DRIFT

AGRICULTURAL

PESTICIDE APPLICATIONS

PESTICIDE & FORMULATION ADDITIVES

SOIL APPLICATION

ADSORPTION & DIFFUSION IN SOIL SOLUTION

ADSORPTION TO MINERAL/ORGANIC SURFACES

UPTAKE, ACCUMULATION, FOOD CHAINS (MICROBES, INVERTEBRATES & VERTEBRATES)

DEATH, FAECES

NON-TARGET ECOLOGICAL EFFECTS

SPRAYS/DUSTS

RUN-OFF

SEED DRESSING

UPTAKE

INTRACELLULAR AND EXTRACELLULAR MICROBIAL METABOLISM

SOIL CHEMICAL TRANSFORMATIONS

LEACHING

METABOLISM

TRANSPORT

DIFFUSION

UPTAKE

ROOT EXUDATION

PLANT ENZYMES

GASEOUS PRODUCTS

WIND BLOWN PLANT DEBRIS, POLLEN, SEEDS etc.

DEATH OF PLANT & SOIL INCORPORATION

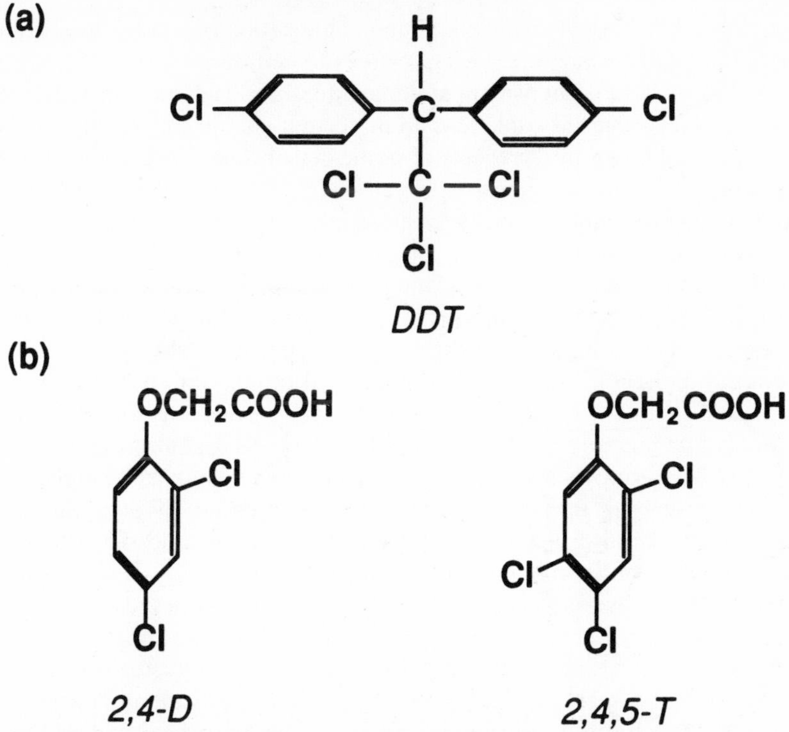

(a)

DDT

(b)

2,4-D   *2,4,5-T*

Figure 6.2 Structure of three pesticides. (*a*) The persistent insecticide DDT (1,1,1-trichloro-2, 2-bis(p-chlorophenyl)-ethane). (*b*) The non-persistent herbicide 2,4-D (2,4-dichlorophenoxy-acetic acid) and the persistent herbicide 2,4,5-T (2,4,5-trichlorophenoxyacetic acid).

different tillage practices may have greater long-term impact on the soil invertebrate community than any direct effects of the pesticides. The introduction of hormone herbicides such as 2,4-D into cereal farming eliminated broad-leaved weeds, thereby allowing the continuous culti-vation of wheat and minimum tillage systems. In this case the use of *herbicides* has indirectly increased the numbers of earthworms in soils under reduced cultivation.

Broad-spectrum biocides including *soil fumigants* often cause an increase in the growth of plants even when no obvious pests or diseases are present. This may involve the release of nitrogen from the pesticide or from the killed biomass (section 2.6.2), or the control of deleterious rhizobacteria

Figure 6.1 Pesticides in the soil environment (*solid lines* movement of pesticides or products, *broken lines* agencies affecting the pesticide or products, *shaded area* rhizosphere, *animation* micro-organisms). (Courtesy of Dr I.R. Hill, ICI.)

(section 6.3.3), perhaps by encouraging colonisation of soil by fluorescent pseudomonads, which often occurs after fumigation.

A change in the input of plant material into soil will have a major effect on all organisms. The rhizosphere is an important site for microbial activity, but little is known of the effects of pesticides on micro-organisms in this zone of soil. Herbicides increase the amount of material lost from the roots, and plants exposed to herbicides may be more susceptible to plant disease than untreated plants. Most herbicides do not penetrate more than 1–2 cm into the soil; side-effects in this zone may be unimportant when there is a large volume of deeper soil in which roots are present and microbial activity is proceeding normally. Herbicides do not appear to have any great or prolonged direct effect on the total bacterial population of the soil.

Despite many years of study it is not yet possible to unequivocally identify and quantify the effects of pesticides on soil organisms and on soil fertility. This is partly due to our lack of complete understanding of the role of organisms in the development and functioning of soil. Suitable techniques are also lacking. For example, the importance of symbiotic nitrogen fixation in maintaining or improving the fertility of soils is well established, however, there are no reliable techniques available for routinely measuring rates of nitrogen fixation by legumes and actinorhizal plants. There is also a need to assess the significance of fluctuations in soil organisms and biological processes caused by a pesticide in relation to natural fluctuations. For example, drying and wetting of soil can lead to a 50% decrease in microbial activity, and organisms may take up to 30 days to recover. Pesticide effects which do not exceed these limits should not therefore present any risks.

6.1.2 Recalcitrant compounds

The magnitude of the side-effects associated with a particular pesticide will depend partly upon the persistence of the compounds in soil. The organo-chlorine insecticide DDT (Figure 6.2a) and the organo-phosphorus insecticide parathion (O,O-diethyl o-p-nitrophenylphosphoro-thioate) persist in soil for more than 15 years, whereas the herbicide 2,4-D (Figure 6.2b) disappears from soil within several weeks. The reasons why some compounds are more resistant to attack by micro-organisms (*recalcitrant*) in soil are complex. Simple structural changes can convert a rapidly decomposable substrate into a persistent compound. For example, the addition of an extra chlorine atom to a molecule of 2,4-D (Figure 6.2b) converts it into 2,4,5-T which persists in soil for several months.

Soils may develop the ability to degrade a pesticide following exposure to the compounds. This apparent adaptation by the soil may involve induction of enzymes by micro-organisms or the transfer of plasmids (section 6.4.2) containing genes responsible for enzymes involved in the breakdown of the compound. For example, the genes responsible for degradation of 2,4-D are carried on a conjugative plasmid in *Alcaligenes paradoxus*. Non-enzymatic reactions such as changes in pH brought about by micro-organisms may also contribute to pesticide degradation. Similarly, recalcitrance may be partly due to adverse physical conditions such as low concentrations of oxygen, or physical protection of the compounds by soil components. It is likely that the cooperative action of communities of micro-organisms is important in the degradation of *xenobiotic* compounds (chemicals that are foreign to the soil).

6.2 Inorganic and organic nitrogen fertilisers

The use of nitrogen fertilisers has increased rapidly during the past 30 years, with most fertiliser nitrogen applied as NH_4^+ containing compounds such as ammonium nitrate or urea. Urea, the most widely used nitrogen fertiliser worldwide, is rapidly hydrolysed to NH_4^+ in soil (section 2.5.3) and then oxidised to NO_3^- (section 2.4.2). Under alkaline conditions, NH_4^+ may be converted into ammonia gas which is lost by *volatilisation* to the atmosphere.

There are increasing concerns that intensive use of nitrogen fertilisers may lead to higher concentrations of NO_3^- in surface and groundwaters, which can cause eutrophication (excessive algal growth and depletion of oxygen as this undergoes decomposition) of lakes and streams. High concentrations of NO_3^- in leafy vegetables and in drinking water may cause health problems. The maximum limit for drinking water recommended by the World Health Organisation is $50\,mg\,NO_3^-\,l^{-1}$, and this is also the maximum peak concentration allowed by the European Commission. Leachate from arable soils in the UK often exceeds this limit.

Studies in the UK have indicated annual rates of *leaching* of nitrogen from an unfertilised soil of $0.2\,g\,N\,m^{-2}$ under grass, $3.0\,g\,N\,m^{-2}$ under white clover and $13.7\,g\,N\,m^{-2}$ under bare fallow. Rates of leaching from fertilised grassland are higher under grazing than when grass is cut. The highest concentrations of NO_3^- in aquifers are found under fertilised arable soils, and large increases in concentrations occur following the ploughing of established grassland.

Data from the Rothamsted Drain Gauges established in 1870 indicate rates of NO_3^- leaching at the end of the nineteenth century from uncropped, uncultivated arable soils receiving no manure or fertiliser which are similar to current rates of leaching from fertilised wheat crops. From this data it can be concluded that the major source of NO_3^- leaching from arable soils is not directly from unused chemical nitrogen fertilisers, but from mineralisation of organic nitrogen (which may have been built up over a long period of time by fertiliser additions). The half-life for this organic nitrogen has been estimated to be 41 years, indicating the long-term nature of the problem. There are no obvious ways of reducing the concentration of NO_3^- in groundwaters in the short-term, other than treatment of the water before consumption.

Apart from leaching, losses of nitrogen occur due to volatilisation of NH_3 and denitrification (section 2.4.3). Emissions of nitrogen from animal slurry produced by intensive livestock enterprises are contributing to acid deposition in countries such as the Netherlands (section 4.3.3). These processes also contribute to the inefficiency of utilisation of nitrogen in both inorganic and organic fertilisers. Control over rates of urea hydrolysis and nitrification in soils offers one approach to reducing the losses of nitrogen from both inorganic and organic nitrogen fertilisers to groundwaters and to the atmosphere.

6.2.1 Inhibitors of nitrogen transformations

Numerous compounds have been patented as inhibitors of urea hydrolysis and of nitrification, but most are not particularly effective. The most interesting urease inhibitor is phenylphosphorodiamidate (PPD) but the activity of this compound is reduced at temperatures above 10 °C. There appears to be scope for selecting similar compounds which are less affected by high temperatures and which do not affect other soil nitrogen transformations.

The control of nitrification appears more promising. Of the numerous patented compounds three have attracted significant attention. These are 2-chloro-6-(trichloromethyl) pyridine (marketed as Nitrapyrin or N-serve), dicyandiamide (marketed as Didin or DCD) and 5-ethoxy-3-(trichloromethyl)-1,2,4-thiadiazole (marketed as Dwell or Terrazole). DCD breaks down in soil to form NH_4^+, NO_3^-, water and carbon dioxide, however, the fate of Nitrapyrin is uncertain and toxic residues may be produced. These compounds are useful in reducing the losses of nitrogen from animal slurries applied to the land during the winter when up to 20%

Table 6.1 The effect of nitrification inhibitors on the response of a range of crops to animal slurry (surface applied (S), or injected (I)). After B.F. Pain *et al.* (1987), in *Animal Manures on Grassland and Fodder Crops. Fertilizer or Waste*, eds. H.G. van der Meer *et al.*, Martinus Nijhoff, Groningen, 229–246.

Crop	Country	Slurry	Inhibitor	Yield (g m^{-2})
Grass	UK	Cattle (S)	– DCD	220*
Grass	UK	Cattle (S)	+ DCD	300*
Maize	USA	Pig (I)	– Nitrapyrin	730
Maize	USA	Pig (I)	+ Nitrapyrin	1140
Sugar beet	N'lands	Pig (S)	– DCD	6200
Sugar beet	N'lands	Pig (S)	+ DCD	7540

*First silage cut only.

of the total nitrogen applied may be lost by denitrification from injected slurry. Table 6.1 shows the effectiveness of Nitrapyrin and DCD in controlling the losses of nitrogen and hence increasing the yield of a range of crops receiving animal slurry.

6.3 Inoculation

If an important organism is absent from a soil then it may be necessary to introduce or *inoculate* that organism into the soil. For example, some trees will not grow unless the appropriate ectomycorrhizas are present, and the rhizobia present in a soil may not nodulate a new legume which has not previously been grown in that soil. Inoculation may be carried out by transferring soil from a region where the plant grows successfully to the new region. As soil microbiological processes have become better understood attempts have been made to inoculate plants with pure cultures (or known mixtures of pure cultures). Such attempts to introduce an organism into a vacant *niche* are generally successful. However, inoculation may also be used in an attempt to supplant an indigenous organism, which is of poor quality, with a superior strain. This is more difficult to achieve.

Azotobacter chrococcum and *Bacillus megatherium* were used as seed inoculants in the USSR and eastern Europe for several decades. This *bacterisation* of crops was aimed at exploiting the nitrogen-fixing and phosphate-solubilising abilities of *A. chrococcum* and *B. megatherium* respectively. Yield increases of up to 10% were achieved with *A. chrococcum*, but the beneficial effects were probably due to the excretion of plant growth hormones (sections 3.2.1 and 3.2.3). The responses to inoculation with *B. megatherium* were less convincing.

Organisms may also be introduced to remove or overcome the effects of detrimental organisms such as a pathogen (biological control). Such uses of biological agents to improve plant growth may be considered more desirable than the use of chemical means, particularly in cases where chemical treatment is not particularly effective. However, as the techniques of molecular biology are applied to the development of microbial inoculants, the acceptability of biological agents is being called into question.

6.3.1   Rhizobium

The introduction of root-nodule bacteria (*Rhizobium sp.* and *Bradyrhizobium sp.*, section 3.4.2) into soils, usually by sticking a carrier substance such as finely-milled peat containing the organisms onto the legume seed, represents the earliest attempt at inoculation. Patents were first taken out on *Rhizobium sp.* in the UK and USA in 1885, and in 1985 the market for soybean inoculants was valued at 18×10^6.

Soybean (*Glycine max*) is an example of a crop which has a specific requirement for its symbiotic partner; it is nodulated only by *Bradyrhizobium japonicum* (formerly *Rhizobium japonicum*) and this species normally only nodulates soybeans. Therefore, if soybeans have not been grown before in a soil the appropriate symbiotic partner will be absent, and if nitrogen is limiting growth, the plant will respond to inoculation. This is the case in Africa where attempts have been made to introduce US soybean cultivars as a source of protein and oils, and in Canada where chickpeas (*Cicer arietinum*) are being increasingly grown (Table 6.2).

Table 6.2 Response of different legumes to inoculation with rhizobia in terms of nodulation (+ present, − absent) and yield (seed or dry matter yield). After E.L. Pulver *et al.* (1982) *Crop Science* **22**, 1065–1070, R.J. Rennie and S. Dubetz (1986) *Agronomy Journal* **78**, 654–660, and P. Newbould *et al.* (1982) *Journal of Agricultural Science Cambridge* **99**, 591–610.

Legume	Country	Uninoculated		Inoculated	
		Nodules	Yield $(\mathrm{g\,m^{-2}})$	Nodules	Yield $(\mathrm{g\,m^{-2}})$
Soybean	Tanzania	−	50	+	177
Chickpea	Canada	−	145	+	224
White clover	Scotland	+	90	+	93

In certain soils the appropriate root nodule bacteria may be present, but they may be poor at fixing nitrogen. This is often the case in acid soils such as the hill and upland areas of the UK. Attempts have been made to introduce improved strains of clover rhizobia (*Rhizobium leguminosarum* biovar *trifolii*, formerly *Rhizobium trifolii*) into these areas, but responses to inoculation by the host legume have been disappointing (Table 6.2).

Attempts to supplant poor indigenous strains by superior strains have generally been unsuccessful. In many areas of the USA where soybeans have been grown for some time one particular strain of *B. japonicum* (serotype 123, probably introduced originally from China or Japan) is the dominant organism forming nodules, and attempts to introduce improved stains have met with little success. We still do not fully understand the factors that determine the ability of an organism to compete successfully in soil with other organisms of the same species or different species (*competitive ability*).

Some legume crops have become widely used throughout the world without the deliberate use of inoculation. Many of the tropical grain legumes such as cowpea (*Vigna unguiculata*) and pigeon pea (*Cajanus cajan*) are nodulated by organisms that will infect and fix nitrogen in a wide range of legumes. The common bean (*Phaseolus vulgaris*) was introduced into east Africa about 350 years ago when trade between Africa and the Americas was started, and forms nodules (although often poorly) without the need for inoculation. There is evidence that the native strains of root nodule bacteria that nodulate some of the indigenous tree legumes such as *Acacia sp.* are able to nodulate the common bean. Little is known about this, but the selection of promiscuous cultivars which are able to form effective nodules with the native root nodule bacteria offers an alternative stategy for the improvement of nitrogen fixation by tropical legumes. For example, locally-bred soybean cultivars in Tanzania nodulate well with indigenous organisms and yield as much as inoculated cultivars bred in the USA.

6.3.2 Biological control of plant root pathogens

Biological control can be defined as the use of one or more organism, including the host plant but excluding man, to decrease the inoculum or the disease-producing activity of a pathogen. It is based upon the antagonistic action of micro-organisms which has been known as a laboratory phenomenon for over a century. The observation that certain soils are *suppressive* to diseases such as *Fusarium* wilt was first made over eighty years ago, but our understanding of suppressiveness remains poor.

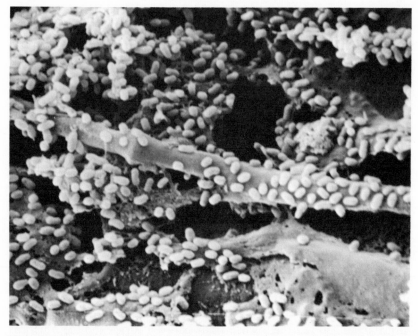

Figure 6.3 The pathogenic fungus *Gaeumannomyces graminis* growing on a root of wheat (*Triticum aestivum*) colonised by the bacterium *Pseudomonas fluorescens* inoculated into soil (magnification × 5300). Courtesy of Dr R. Campbell, University of Bristol.

A decline in the severity of take-all disease of wheat (*Triticum aestivum*), caused by *Gaeumannomyces graminis*, is often observed when wheat is grown continuously in a soil. This decline has been associated with an increase in the population of fluorescent pseudomonads which produce antibiotics when grown in laboratory media. It has been proposed that these bacteria are antagonistic to the causative organism of take-all, possibly by colonising the root surface preferentially, or by producing antibiotics. Figure 6.3 shows a wheat root infected with take-all also inoculated with a strain of *Pseudomonas fluorescens*.

Attempts have been made to isolate these organisms for use as inoculants to reduce the severity of take-all disease. In the glasshouse, wheat plants often respond favourably to inoculation with fluorescent pseudomonads in soils infected with take-all, but under field conditions responses are less impressive and are variable from year to year. Table 6.3 shows the seed yield for wheat inoculated with a mixture of two strains of fluorescent

Table 6.3 Response of wheat (*Triticum aestivum*) to inoculation with a mixture of two strains of *Pseudomonas fluorescens* in a soil in the field with and without take-all disease (*Gaeumannomyces graminis*), in the USA. After D.M. Weller and R.J. Cook (1983) *Phytopathology* **73**, 463–469.

Take-all present	Inoculated with pseudomonads	Seed yield (g m^{-2})
+	−	325
+	+	413
−	−	571

pseudomonads in soils which were infected with take-all. The bacterial inoculant gave a significantly higher yield when take-all was present, but the yield was still much less than in the complete absence of the disease. It remains to be seen whether or not a single antagonistic organism can provide as effective control of a disease as the total microflora of the suppressive soil from which it was isolated.

Effective plant disease control is more likely to be achieved using a series of complementary procedures, integrating chemical and biological methods. An example of *integrated pest control* comes from the Californian citrus orchards, where carbon disulphide has been used since 1914 to control root rot due to the fungus *Armillaria mellea*. This works by making the fungal mycelium more susceptible to another fungus *Trichoderma viride* which in turn controls the disease. Organisms are now being selected for resistance to fungicides which will allow the use of both biological and chemical control agents.

6.3.3 Plant growth-promoting rhizobacteria

It has been estimated that 8–15% of the bacteria isolated from the rhizosphere are deleterious to the plant, and 2–5% are beneficial. Although it is difficult to balance the advantages and disadvantages of different organisms (section 3.2), it is clear that the efficiency of roots and plant growth may be adversely affected by mild pathogens in the rhizosphere producing compounds which inhibit root growth and activity. Micro-organisms also stimulate root exudation which diverts photosynthate from other uses within the plant. Other rhizosphere inhabitants may inhibit or displace these organisms or overcome their effects by producing antibiotics or plant growth-promoting substances.

Table 6.4 Effect of inoculation with a wild-type sidero-phore-producing (sid⁺) strain and non-siderophore-producing mutant (sid⁻) strains of *Pseudomonas fluorescens* (PGPR) on yield (fresh weight) of seed potatoes in the field 86 days after planting, in the Netherlands. After B. Schippers *et al.* (1987).

Treatment	Tuber yield $(\mathrm{g\,m^{-2}})$
No inoculant	3340
WCS358 (sid⁺)	3720
WCS358 (sid⁻)	3390

An increase in the population of pseudomonads inhibitory to the growth of wheat (*T. aestivum*) roots has been observed in minimum cultivation systems compared to conventional systems. These organisms have been implicated in the decreased cereal yields that are often associated with monocultures in both the USA and in England. In the Netherlands, the yield of potatoes (*Solanum tuberosum*) declines as the frequency of potato cultivation within a rotation increases. The production of *cyanide* by members of the rhizosphere microflora has been suggested as a cause of this, inhibiting root cell energy metabolism and nutrient uptake. The introduction of specific root-colonising fluorescent pseudomonads into soils showing this yield reduction can increase the yield of potatoes (Table 6.4).

A hypothesis has been developed to explain these observations. *Siderophores*, which are produced by the plant growth-promoting rhizobacteria (PGPR) under conditions of iron limitation, compete for iron in the rhizosphere with those siderophores produced by the deleterious bacteria. The higher affinity for iron by the PGPR siderophore leads to inhibition of the deleterious organisms. Support for this hypothesis is provided by the observation that a transposon-induced mutant (section 6.4) which no longer produced siderophores (sid⁻), but which colonised the roots of potato did not improve yields of potato (Table 6.4).

Inoculation of tubers with PGPR only increases the yield of seed potatoes not that of ware potatoes harvested 1–2 months later. Root exudates change as plants age and proline, which stimulates cyanide production, increases in potatoes that are stressed. The deleterious microbial population may, therefore, become more active later in the season. Responses to inoculation with PGPR in the field are often variable. Poor colonisation of the root is the major factor limiting commercial exploitation of these organisms, but the factors which determine the ability of an organism to successfully colonise the rhizosphere and rhizoplane are still not known.

6.3.4 *Mycorrhizas*

There has been much interest in the inoculation of crop plants and trees with mycorrhizal fungi (section 3.7). Encouraging responses by trees to inoculation with ectomycorrhizas have been observed, for example in the afforestation of treeless grassland areas of the USA, the steppes of the USSR and severely disturbed sites such as mine spoil. Ectomycorrhizal fungi can be grown in pure culture and this assists greatly in the production of inoculants. However, it is not known what properties are required for an efficient symbiosis, nor what genetic variation exists and what breeding systems operate in these fungi. The inoculation of pine seedlings in the nursery with a pure culture of *Pisolithus tinctorius* resulted in improved establishment in the nursery and increased survival and growth of seedlings when transplanted to mine spoil with a history of repeated failures of pine plantations.

It is not yet possible to grow VAM fungi in pure culture. Although inoculation, which usually involves the use of chopped infected plant material, has shown some promise in pots in the glasshouse, responses to inoculation in the field have been disappointing.

6.3.5 *Agrobacterium rhizogenes* strain 84

In 1972 a non-pathogenic strain of *Agrobacterium rhizogenes* (strain 84) was discovered which, when inoculated onto plants in soil, decreased the incidence of crown galls on peach. Since that time, this single strain has been used worldwide as a biological control agent. Seeds, cuttings or roots of young plants are dipped in a suspension of the organisms and almost 100% control of the disease is achieved. There are some cases where inoculation has been unsuccessful, for example, crown gall was not prevented on grapes in Greece.

Strains of *A. tumefaciens*, the causative agent of crown galls (section 3.6), can be grouped according to the type of opines synthesised by the transformed plant; octopine, nopaline, or agropine. The nopaline strains are responsible for the most economically serious damage. Strain 84, which is only effective against nopaline strains, can metabolise nopaline and competes with the pathogen for this substrate, and also produces agrocin 84, a bacteriocin which kills the pathogen. Agrocin 84 is taken up by the pathogen by an uptake permease system which is normally used to assimilate agrocinopine, another opine produced by plant cells transformed by nopaline strains.

The genes for agrocin 84 are carried on a small conjugative plasmid (section 6.4), and transconjugants can synthesise the bacteriocin. If the recipient is a pathogenic strain then the transconjugant retains its pathogenicity, produces agrocin 84 and is also resistant to the bacteriocin. This could pose serious problems because the pathogen is then no longer under biological control. The transfer region of the plasmid has been located by transposon mutagenesis, and a deletion mutant constructed (section 6.4). This should prevent transfer of T-DNA to other strains, and prolong the usefulness of this biological control agent. Such a genetically engineered micro-organism is likely to be scrutinised carefully before its use is allowed in the field.

6.4   Use of genetically engineered micro-organisms

The genetic material in a bacterium is contained in the *chromosome*, a large circular piece of DNA containing approximately 1000 kb (the average gene contains 900 base pairs, 1 kb equals 1000 base pairs, and 1 MDa equals approximately 1.5 kb), and *plasmids* which are small circular pieces of extrachromosomal DNA containing 1–300 kb. Plasmids, which can carry a substantial proportion of a cell's genetic material and which may be present in multiple copies (up to 40), replicate independently of the chromosome and often carry genes which assist with survival under adverse conditions. For example, resistance to antibiotics, heavy metals and bacteriocins (section 6.3.5), and catabolism of certain sugars and xenobiotics (section 6.1.2), together with the ability to form root nodules and crown galls (Chapter 3) are all determined by plasmid-borne genes.

Transposon mutagenesis is often used to locate particular genes in a bacterium. A transposon is a highly mobile short piece of DNA, often containing antibiotic-resistance genes as markers, which integrates itself at random into the chromosome or plasmid, and in the process deletes genes and causes a mutation. Mutants are selected which have lost the particular gene(s) of interest and the position of the transposon and hence the original gene(s) can be located using DNA hybridisation techniques.

There are three main ways in which DNA may be transferred between bacteria of the same or different species. *Conjugation*, which only occurs in Gram-negative bacteria, involves conjugative plasmids (a minority of plasmids) and requires cell-to-cell contact between donor and recipient during which the plasmid and all or part of the chromosome is transferred. In *transduction* a temperate bacteriophage (a virus) adsorbs to the surface of a bacterium and injects its viral DNA into the bacterial cell where it may integrate into the chromosome or exist as a plasmid. Death of the bacterial cell does not occur until the cell is exposed to agents such as ultraviolet light which induce the cell to produce intact phage particles which may contain additional DNA from the bacterium. The cell then lyses and the phages are released and may infect other bacteria. *Transformation* occurs under adverse conditions when bacteria can take up naked DNA which becomes integrated into the host chromosome.

Recombinant DNA technology offers the possibility of improving microbial inoculants for use in soil. When suitable genes have been identified they can be introduced into a *vector* such as a plasmid or a bacteriophage. This involves cutting and joining the vector DNA. The genes are usually transferred into *Escherichia coli* and amplified (*cloned*), before being transferred to the desired host organism. Some progress has already been made, for example *Agrobacterium rhizogenes* strain 84 (section 6.5.3). Strains of *Rhizobium sp.* have been constructed with altered host range by removal (*curing*) and introduction of plasmids (Table 6.5) and with introduced antibiotic-resistance marker genes. Genes have been identified in *Pseudomonas fluorescens* which determine the production of a phenazine-type antibiotic which appears to be essential for the suppression of take-all disease (section 6.3.2).

The chitinase genes from *Serratia marcescens* have been cloned in *E. coli* and transferred to *P. fluorescens*, an efficient root-colonising organism, and it is hoped that plants inoculated with this organism will be protected

Table 6.5 Nodulating ability of *Rhizobium leguminosarum* transconjugants produced by Sym plasmid transfer. After M.A. Djordjevic *et al.* (1983) *Journal of Bacteriology* **156**, 1035–1045.

Strain	Phenotype	Nodulation response on white clover	pea
ANU843	Wild type	Nod +	Nod −
ANU845	Sym plasmid cured	Nod −	Nod −
ANU845(pBRIAN)	Transconjugant	Nod +	Nod −
ANU845(pJB5JI)	Transconjugant	Nod −	Nod +
ANU300	Wild type	Nod −	Nod +
ANU615	Sym plasmid cured	Nod −	Nod −
ANU615(pBRIAN)	Transconjugant	Nod +	Nod −
ANU615(pJB5JI)	Transconjugant	Nod −	Nod +

against fungi with chitinous cell walls and possibly soil-inhabiting insects. *Bacillus thuringiensis* is another organism of interest because it produces large insecticidal proteinaceous crystals which are lethal to Lepidoptera, Coleoptera and Diptera. Spores and crystal preparations from *B. thuringiensis* have been used for almost 35 years to control Lepidoptera. The toxins are attractive because they are active in low doses and are highly specific. The genes which allow production of these toxins have been inserted into the chromosome of a root-colonising fluorescent pseudomonad with the aim of using this organism to control cutworm.

Micro-organisms in soil and water exhibit a wide range of degradative abilities; however, in order to exploit these abilities it may be necessary to accelerate the process or combine the abilities of several organisms into a single strain. Many xenobiotics are recalcitrant (section 6.1.2) and it is therefore desirable to develop new organisms to degrade them. This may involve the modification of an existing pathway for a similar compound, or the design of a new pathway. Micro-organisms have been constructed which are capable of degrading chemical pollutants such as 2,4,5-T.

6.4.1 *Fate of non-engineered organisms in soil*

There is a large body of information on the fate of non-engineered micro-organisms, in particular *Rhizobium sp.* and *Bradyrhizobium sp.*, introduced into soil. *Rhizobium* inoculants have been used for almost 100 years and millions of hectares are inoculated annually. There have been no unforeseen effects associated with this massive soil inoculation programme, despite the fact that many of the inoculants used may have been contaminated with organisms other than *Rhizobium*. On the contrary, the evidence from *Rhizobium* inoculation suggests that it is extremely unlikely that an introduced organism will persist in soil unless there is a vacant niche for the introduced strain (section 6.3.1).

One of the organisms of interest as a soil inoculant is *Pseudomonas sp.*, a genus which is similar in many respects to *Rhizobium*. Studies on *Rhizobium* have indicated that survival of inoculants in soil depends upon a wide range of soil factors such as acidity, temperature, moisture content, predation and antagonism from other soil organisms (Chapter 4). The response of *Pseudomonas fluorescens* to acidity and related factors is similar to that of *Rhizobium*, therefore the information available on root nodule bacteria should be valuable in predicting the fate of pseudomonads in soil.

Large numbers of micro-organisms are introduced into soil when animal slurry is applied to the land. The microbial population in slurry is variable, but contains a proportion (approximately 1%) of coliform bacteria, which are normally found in the guts of animals and some of which may be pathogenic to humans. Coliforms do not usually survive more than a few weeks in soil, but this could be long enough to allow genetic transfer to occur. *Escherichia coli*, a coliform bacterium, is widely used in recombinant DNA studies (section 6.4). It can transfer genes to about 40 genera of Gram-negative bacteria including some pathogens. In this type of work strains which are unable to colonise the human gut, and which are unable to survive in soil are used.

6.4.2 *Genetic exchange in soil*

Transfer of genetic material between bacteria occurs in soil. Conjugation has been reported in sterile and non-sterile soil for *E. coli* and *P. aeruginosa*, transformation in sterile soil for *Bacillus subtilis*, and transduction in sterile soil for *E. coli*. The optimum temperature for conjugation is greater than 20 °C and the optimum pH value greater than 6. Bacteria may lose plasmids at high temperatures, for example *Rhizobium sp.* lose the Sym plasmid when incubated at 37 °C. Novel genes should be more stable if located on the chromosome rather than on plasmids. Sites of high population density in soil, such as the rhizosphere and invertebrate guts, may provide bacteria with greater opportunities for genetic exchange.

Subtle differences in properties such as antibiotic resistance and metabolic capacity between donor and recipient *transconjugants* may influence competitive ability of these organisms in soil, although this has not been demonstrated. There is some evidence that transconjugants formed in the laboratory are less able to survive in soil than the parent organisms. However, other evidence indicates no change in properties such as root-colonising ability.

6.4.3 *Risk assessment*

The use of *genetically engineered micro-organisms* (GEMs) in soil, and the accidental release of GEMs into soil, raises several important questions which need to be considered before permission is given by the appropriate legislative bodies for the use of such an organism. Can the organism survive, multiply and migrate in soil? Is the novel gene transferred to other organisms? Does the GEM affect other soil processes? What are the overall risks associated with the use of the organism?

The answers to these questions are not readily available. In order to predict the likely behaviour of an introduced organism in soil an understanding is required of the biology of the soil. Although we have some information on certain organisms in soil, and on how they interact with other organisms and with the soil, we are still far from a complete understanding. We still do not know what properties confer on a bacterium the ability to compete successfully with other soil organisms, or to colonise the surface of a root. At present it is impossible to predict with any degree of certainty the fate of an introduced organism in soil or to design a successful inoculant for use in the field. This is largely due to the problems in knowing the precise nature of the ecological niche into which the organism is to be introduced (Chapter 4).

It is necessary therefore to carry out a case-by-case assessment of the risks associated with the use of a particular GEM. It is essential that reliable methods are available for detecting an introduced organism in soil. A range of techniques are available including traditional ecological methods such as culturing on selective media and fluorescent antibody techniques, which allow the *phenotype* of the organism to be monitored. Recently developed techniques such as ^{32}P-labelled DNA probes, which only hybridise (attach) to complementary sequences of DNA, allow the *genotype* of the organism to be monitored.

Risks can be assessed in model soil systems in the laboratory (*microcosms*) in which field conditions can be simulated as closely as possible under strict containment. It may be desirable to simplify the system, for example by sterilising the soil, or by using sieved soil, however, the extrapolation of results from microcosms to the field becomes more difficult as microcosms become less representative of field conditions.

6.5 Sustainable agricultural systems

The remarkable increase in food production achieved in developed countries during the last 30 years, and more recently in some developing countries such as India, has been due to the breeding of high-yielding varieties of crops such as wheat, increasing mechanisation and the use of large quantities of nitrogen and phosphorus fertilisers and pesticides. However, surplus food production in developed countries together with increasing concern about the effects of agricultural practices on the environment are stimulating a critical evaluation of the long-term feasibility of high-input agriculture.

At the same time, efforts are continuing to improve food production in developing countries. Many of the improvements which contributed to the *green revolution* were based on improved varieties of rice which responded well to inputs of fertiliser and pesticides. This was of benefit to the larger, more mechanised farmer, but did not help the small farmers. For example, only a small proportion of farmers in south-east Asia have irrigation facilities and are able to take full advantage of the improvements in rice production by growing up to four crops per year. The majority of farmers cannot afford to buy fertilisers and are restricted to growing only one crop per year.

The traditional *shifting cultivation* (slash and burn) practised widely in the humid and sub-humid tropics allows sustainable food production at low population densities. Following clearance of the forest, soil nutrients are depleted during a short cropping cycle and then restored during a fallow period. However, increasing population pressure has led to a decrease in the length of fallow period with a consequent decrease in fertility and crop yields. Attention is therefore being focused on alternative low-input sustainable systems.

6.5.1 *Mixed cropping*

Mixed cropping is commonly practised in the tropics, often as an insurance against failure of one of the crops. This may involve *intercropping* (growing two or more crops simultaneously) or *sequential cropping* (growing two or more crops in sequence). Benefits are particularly pronounced when one of the crops is a legume. For example, the grain yield of a non-legume following a grain legume is often 50–100% greater than the yield of the same crop grown as a monoculture, equivalent to a fertiliser application of 5–10 g N m^{-2}.

The main contribution of the legume is its ability to fix atmospheric nitrogen (section 3.4), thereby sparing some soil nitrogen for the non-legume. It is less likely that legumes provide a net input of fixed nitrogen into the soil because the removal of nitrogen in the seed is often greater than estimated rates of nitrogen fixation. Other benefits of having a legume in a mixed cropping system include improvements in soil structure, disruption of pest and disease cycles and an input of crop residues with a C:N ratio which are more likely to result in net mineralisation of nitrogen during the early stages of decomposition than are crop residues from a non-legume (section 2.2.1).

The use of a woody species as a component of a mixed cropping system (*agroforestry*) is attractive because trees, particularly leguminous species,

Table 6.6   Nitrogen yield from hedgerow prunings and the yield and nitrogen uptake of maize (*Zea mays*) in an alley cropping system in Nigeria. After B.T. Kang (1988).

Hedgerow species	Prunings N yield ($g\,N\,m^{-2}$)	Maize N yield ($g\,N\,m^{-2}$)	Maize grain yield ($g\,m^{-2}$)
None	None	2.6	163.2
Alchornea cordifolia (non-legume)	6.2	4.5	255.7
Leucaena leucocephala (legume)	23.1	6.8	321.0

can improve recycling of nutrients and water from deep in the soil, provide shade and supply food, firewood and wood products. Most attention has been focused on *Leucaena leucocephala*, but other tree legumes such as *Sesbania sp., Prosopis sp.* and *Cassia sp.* may have greater potential. A system which uses pruned hedgerows of woody plants intercropped with food crops (*alley cropping*) has been recently developed. Table 6.6 shows the increase in yield of maize in an alley-cropping system in West Africa using either a non-leguminous or a leguminous tree.

Sustainable yields of crops have been achieved in the traditional *home gardens* and *village forest gardens* of Malaysia, Indonesia and the West Indies. These systems simulate the forest environment and have a multi-storeyed structure and a high diversity of cultivated species. On the island of Java mixed gardens have been in use since at least the tenth century and are still widespread in areas of high population density. A typical garden might include a ground layer of vegetables such as beans and tomatoes, a storey 1.5–5 m high of food plants such as cassava and papaya, and upper storeys with a range of different tree heights, for example citrus, coffee, mango, bamboo and coconuts extending up to 35 m. The complex interactions that occur between such a mixture of plant species are worthy of study.

6.5.2   Organic farming

No modern agricultural systems are completely self-contained, and all have to rely on some form of nutrient input if yields are to be sustained in the long term. The provision of an input of nitrogen into soils from nitrogen fixation forms an important part of all low-input farming systems. The use of legumes such as white clover (*Trifolium repens*) in developed countries has declined as the use of nitrogen fertiliser has increased. However, the fixation

Table 6.7 Yield of unfertilised ryegrass (*Lolium perenne*) grown as a monoculture or in a mixture with white clover (*Trifolium repens*) in the UK. A.M. McNeill and M. Wood, unpublished.

Species	Mixture $(g\,m^{-2}\,yr^{-1})$	Monoculture $(g\,m^{-2}\,yr^{-1})$
Ryegrass	485	431
White clover	477	None
Total	962	431

of atmospheric nitrogen by legumes (section 3.4) offers an alternative nitrogen source for soils which may be more sustainable than nitrogen fertiliser.

When effectively nodulated, pasture legumes may increase the nitrogen status of the soil when plant residues are returned to the soil and decomposed, particularly by the grazing animal. Table 6.7 shows the high combined yield of ryegrass and white clover when grown in a mixture compared to the low yield of a grass monoculture, under cutting rather than grazing.

Many of the current objectives of modern agricultural research are compatible with the aims of those involved in *organic farming* (or *biological agriculture*). Biological agriculture aspires to a system in which the maintenance of soil fertility and the control of pests and disease are achieved by the enhancement of natural processes and cycles, with only moderate inputs of energy and resources, while maintaining reasonable productivity. The *biodynamic agriculture* movement takes this approach a step further in trying to develop a complete system of farming that includes ecological, economic and social aspects.

The use of organic material helps to maintain levels of soil organic matter, to improve soil structure (section 5.4.1) and provides a source of nutrients for crop growth. Earthworm populations are also increased. For example, in the long-term experiments at Rothamsted, the Broadbalk plots which have received farmyard manure $(3500\,g\,m^{-2}\,yr^{-1})$ since 1843 have earthworm populations of $89\,m^{-2}$ compared to populations of $6\,m^{-2}$ in the plots receiving no organic manure or fertiliser.

It has been estimated that the amounts of nitrogen, potassium and phosphorus in farmyard manure and animal slurry produced in the UK are equivalent to 37, 65 and 97% respectively of the amounts of those elements applied as inorganic fertilisers. Organic manures are variable in composition and responses by crops are often variable. However, as our understanding of the biology of soil develops our ability to utilise natural resources more efficiently should improve. Detailed long-term studies are required comparing organic or traditional systems with conventional or modern systems.

6.6 Conclusions

Much work remains to be done to improve our understanding of the biology of soil. The development of food production systems which exploit biological processes to a greater extent is now the common aim of scientists in both developed and developing countries. However, the use of pesticides and inorganic fertilisers is likely to continue, combined with biological control agents to form integrated systems of pest and disease control. Although no major adverse consequences from using chemical or biological pesticides have been observed to date, generalisations about lack of effects in the past should not lead to the complacent assumption that no harmful effects will occur in the future. Any attempts to manipulate or manage soil biological processes must be based upon an understanding of soil organisms, their ecology and the processes they carry out, if progress is to be made beyond the current empirical approach.

FURTHER READING

Chapter 1

Brady, N.C. (1984) *The Nature and Properties of Soils*, Macmillan Publishing Company, New York.

Clarholm, M. (1984) Heterotrophic, free-living protozoas: neglected microorganisms with an important task in regulating bacterial populations. In *Current Perspectives in Microbial Ecology*, eds. Klug, M.J. and Reddy, C.A., American Society for Microbiology, 321–326.

Clark, F.E. (1967) Bacteria in soil. In *Soil Biology*, eds. Burges, A. and Raw, F., Academic Press, New York, 15–49.

Duboise, S.M., Moore, B.E., Sorber, C.A. and Sagik, B.P. (1979) Viruses in soil systems. *CRC Critical Reviews in Microbiology* 7, 245–285.

Edwards, C.A. and Lofty, J.R. (1977) *Biology of Earthworms*, Chapman and Hall, London.

Freckman, D.W. and Caswell, E.P. (1985) The ecology of nematodes in agroecosystems. *Annual Review of Phytopathology* 23, 275–296.

Jenkinson, D.S. and Ladd, J.N. (1981) Microbial biomass in soil: measurement and turnover In *Soil Biochemistry* Volume 5, eds. Paul E.A. and Ladd J.N., Marcel Dekker, New York, 415–471.

Moore, J.C., Walter, D.E. and Hunt, H.W. (1988) Arthropod regulation of micro- and mesobiota in below-ground detrital food webs. *Annual Review of Entomology* 33, 419–439.

Russell, E.J. (1957) *The World of the Soil*, Collins, London.

Swift, M.J., Heal, O.W. and Anderson, J.M. (1979) *Decomposition in Terrestrial Ecosystems*, Blackwell Scientific Publications, Oxford.

Warcup, J.H. (1967) Soil fungi. In *Soil Biology*, eds. Burges, A. and Raw, F., Academic Press, New York, 51–110.

White, R.E. (1987) *Introduction to the Principles and Practice of Soil Science*, Blackwell Scientific Publications, Oxford.

Wild, A. (1988) *Russell's Soil Conditions and Plant Growth*, Longmans, London.

Chapter 2

Alexander, M. (1977) *Introduction to Soil Microbiology*, John Wiley and Sons, New York.

Burns, R.G. (1977) *Soil Enzymes*, Academic Press, New York.

Cole, J.A. and Ferguson, S.J. (1988) *The Nitrogen and Sulphur Cycles*, Cambridge University Press.

Jenkinson, D.S. (1981) The fate of plant and animal residues in soil. In *The Chemistry of Soil Processes*, ed. Greenland, D.J., John Wiley and Sons, New York, 505–561.

Jenkinson, D.S. and Ladd, J.N. (1981) Microbial biomass in soil: measurement and turnover. In *Soil Biochemistry* Volume 5, eds. Paul, E.A. and Ladd, J.N., Marcel Dekker, New York, 415–471.

Sprent, J.I. (1987) *The Ecology of the Nitrogen Cycle*, Cambridge University Press.

Stanier, R.Y., Ingraham, J.L., Wheelis, M.L. and Painter, R.P. (1987) *General Microbiology*, Macmillan Education, Basingstoke.
Swift, M.J. and Boddy, L. (1984) Animal–microbial interactions in wood decomposition. In *Invertebrate–Microbial Interactions*, eds. Anderson, J.M., Rayner, A.D.M. and Walton, D.W.H., Cambridge University Press, 89–131.
Williams, S.T. (1985) Oligotrophy in soil: fact or fiction? In *Bacteria in their Natural Environments*, eds. Fletcher, M. and Floodgate, G.D., Academic Press, New York, 81–110.

Chapter 3

Atkins, C.A. (1984) Efficiencies and inefficiencies in the legume/Rhizobium symbiosis—a review. *Plant and Soil* 82, 273–284.
Bauer, W.D. (1981) Infection of legumes by rhizobia. *Annual Review of Plant Physiology* 32, 407–449.
Bowen, G.D. and Rovira, A.D. (1976) Microbial colonisation of plant roots. *Annual Review of Phytopathology* 14, 121–144.
Giller, K.E. and Day, J.M. (1985) Nitrogen fixation in the rhizosphere: significance in natural and agricultural systems. In *Ecological Interactions in Soil*, eds. Fitter, A.H., Atkinson, D., Read, D.J. and Usher, M.B., Blackwell Scientific Publications, Oxford, 127–147.
Harley, J.L. and Smith, S.E. (1983) *Mycorrhizal Symbiosis*. Academic Press, New York.
Kerr, A. (1987) The impact of molecular genetics on plant pathology. *Annual Review of Phytopathology* 25, 87–110.
Lynch, J.M. (1976) Products of soil microorganisms in relation to plant growth. *CRC Critical Reviews in Microbiology* 5, 67–107.
Mai, W.F. and Abawi, G.S. (1987) Interactions among root-knot nematodes and Fusarium wilt fungi. *Annual Review of Phytopathology* 25, 317–338.
Rice, E.L. (1984) *Allelopathy*, Academic Press, New York.
Rolfe, B.G. and Gresshoff, P.M. (1988) Genetic analysis of legume root nodule initiation. *Annual Review of Plant Physiology and Plant Molecular Biology* 39, 297–319.
Rovira, A.D., Foster, R.C. and Martin, J.K. (1979) Origin, nature and nomenclature of the organic materials in the rhizosphere. In *The Soil–Root Interface*, eds. Harley, J.L. and Scott-Russell, R., Academic Press, New York, 1–4.
Schmidt, E.L. (1979) Initiation of plant root–microbe interactions. *Annual Review of Microbiology* 33, 355–376.
Tjepkema, J.D., Schwintzer, C.R. and Benson, D.R. (1986) Physiology of actinorhizal nodules. *Annual Review of Plant Physiology* 37, 209–232.

Chapter 4

Blank, L.W., Roberts, T.M. and Skeffington, R.A. (1988) New perspectives in forest decline. *Nature* 336, 27–30.
Cawse, P.A. (1975) Microbiology and biochemistry of irradiated soils. In *Soil Biochemistry* Vol. 3, eds. Paul, E.A. and McLaren, A.D., Marcel Dekker, New York, 213–267.
Griffin, D.M. (1972) *Ecology of Soil Fungi*, Chapman and Hall, London.
Hang, A. (1984) Molecular aspects of aluminium toxicity. *CRC Critical Reviews in Plant Sciences* 1, 345–373.
Hopkin, S.P. (1989) *Ecophysiology of Metals in Terrestrial Invertebrates*, Elsevier Applied Science, London.
McGrath, S.P., Brookes, P.C. and Giller, K.E. (1988) Effects of potentially toxic metals in soil derived from past applications of sewage sludge on nitrogen fixation by *Trifolium repens* L.. *Soil Biology and Biochemistry* 20, 415–424.
Nye, P.H. (1981) Changes in soil pH across the rhizosphere. *Plant and Soil* 61, 7–26.
Nye, P.H. and Tinker, P.B. (1977) *Solute Movement in the Soil–Plant System*, Blackwell Scientific Publications, Oxford.

Passioura, J.B. (1988) Water transport in and to roots. *Annual Review of Plant Physiology and Plant Molecular Biology* **39**, 245–265.

Rowbury, R.J., Armitage, J.P. and King, C. (1983) Movement, taxes and cellular interactions in the response of microorganisms to the natural environment. In *Microbes in their Natural Environments*, eds. Slater, J.H., Whittenbury, R. and Wimpenny, J.W.T., Cambridge University Press, 299–350.

Stotzky, G. (1986) Influence of soil mineral colloids on metabolic processes, growth, adhesion, and ecology of microbes and viruses. In *Interactions of Soil Minerals with Natural Organics and Microbes*, eds. Huang, P.M. and Schnitzer, M., Soil Science Society of America, 305–428.

Vreeland, R.H. (1987) Mechanisms of halotolerance in microorganisms. *CRC Critical Reviews in Microbiology* **14**, 311–356.

Wallace, H.R. (1968) The dynamics of nematode movement. *Annual Review of Phytopathology* **6**, 91–114.

Chapter 5

Berthelin, J. (1983) Microbial weathering. In *Microbial Geochemistry*, ed. Krumbein, W.E., Blackwell Scientific Publications, Oxford, 223–262.

Darwin, C. (1881) The formation of vegetable mould through the action of worms. In *The Essential Darwin*, ed. Ridley, M., Allen and Unwin, London, 237–256.

Edwards, C.A. and Lofty, J.R. (1977) *Biology of Earthworms*, Chapman and Hall, London.

Grainger, A. (1982) *Desertification*, International Institute for Environment and Development, London.

Haynes, R.J. (1983) Soil acidification induced by leguminous crops. *Grass and Forage Science* **38**, 1–11.

Jenkinson, D.S. (1981) The fate of plant and animal residues in soil. In *The Chemistry of Soil Processes*, ed. Greenland, D.J., John Wiley and Sons, Chichester, 505–561.

Jenny, H. (1980) *The Soil Resource: Origin and Behaviour*, Springer-Verlag, Berlin.

Lynch, J.M. and Bragg, E. (1985) Microorganisms and soil aggregate stability. In *Advances in Soil Science* Volume 2, ed. Stewart, B.A., Springer-Verlag, Berlin, 133–171.

Miles, J. (1985) The pedogenic effects of different species and vegetation types and the implications of succession. *Journal of Soil Science* **36**, 571–584.

Satchell, J.E. (1983) *Earthworm Ecology from Darwin to Vermiculture*, Chapman and Hall, London.

Schatz, A. (1963) Soil microorganisms and soil chelation. The pedogenic action of lichens and lichen acids. *Journal of Agricultural and Food Chemistry* **11**, 112–118.

Simonson, R.W. (1959) Outline of a generalised theory of soil genesis. *Soil Science Society of America Proceedings* **23**, 152–156.

Wood, T.G. (1976) The role of termites (Isoptera) in decomposition processes. In *The Role of Terrestrial and Aquatic Organisms in Decomposition Processes*, eds. Anderson, J.M. and Macfadyen, A., Blackwell Scientific Publications, Oxford, 145–168.

Chapter 6

Addiscott, T.M. (1988) Long-term leakage of nitrate from bare unmanured soil. *Soil Use and Management* **4**, 91–95.

Arden-Clarke, C. and Hodges, R.D. (1988) The environmental effects of conventional and organic/biological farming systems. II Soil ecology, soil fertility and nutrient cycling. *Biological Agriculture and Husbandry* **5**, 223–287.

Hassall, K.A. (1982) *The Chemistry of Pesticides*, Macmillan, London.

Huxley, P.A. (1983) *Plant Research and Agroforestry*, International Council for Research in Agroforestry, Nairobi.

Kang, B.T. (1988) Nitrogen cycling in multiple cropping systems. In *Advances in Nitrogen*

Cycling in Agricultural Ecosystems, ed. Wilson, J.R., Commonwealth Agricultural Bureau, Wallingford, England, 333–348.

Kerr, A. (1987) The impact of molecular genetics on plant pathology. *Annual Review of Phytopathology* **25**, 87–110.

Lynch, J.M. (1983) *Soil Biotechnology*, Blackwell Scientific Publications, Oxford.

Nye, P.H. and Greenland, D.J. (1960) The Soil under Shifting Cultivation, Commonwealth Agriculture Bureau, Wallingford, England.

Schippers, B., Bakker, A.W. and Bakker, P.A.H.M. (1987) Interactions of deleterious and beneficial rhizosphere microorganisms and the effect of cropping practices. *Annual Review of Phytopathology* **25**, 339–358.

Somerville, L. and Greaves, M.P. (1987) *Pesticide Effects on Soil Microflora*, Taylor and Francis, London.

Sussman, M., Collins, C.H., Skinner, F.A. and Stewart-Tull, D.E. (1988) *The Release of Genetically-Engineered Micro-Organisms*, Academic Press, London.

Stotzky, G. and Babich, H. (1986) Survival of, and genetic transfer by, genetically-engineered bacteria in natural environments. *Advances in Applied Microbiology* **31**, 93–138.

Index